SpringerBriefs in Systems Biology

For further volumes:
http://www.springer.com/series/10426

Sara El-Metwally • Osama M. Ouda
Mohamed Helmy

Next Generation Sequencing Technologies and Challenges in Sequence Assembly

Springer

Sara El-Metwally
Department of Computer Science
Mansoura University
Mansoura, Egypt

Mohamed Helmy
Botany Department
and Biotechnology Department
Al-Azhar University
Cairo, Egypt

The Donnelly Centre for Cellular
and Biomolecular Research
University of Toronto (UofT)
Toronto, Canada

Osama M. Ouda
Department of Computer Science
Mansoura University
Mansoura, Egypt

Department of Information Technology
Michigan State University (MSU)
East Lansing, MI, USA

ISSN 2193-4746   ISSN 2193-4754 (electronic)
ISBN 978-1-4939-0714-4   ISBN 978-1-4939-0715-1 (eBook)
DOI 10.1007/978-1-4939-0715-1
Springer New York Heidelberg Dordrecht London

Library of Congress Control Number: 2014934694

Printed on acid-free paper

Springer is part of Springer Science+Business Media (www.springer.com)

Preface

The introduction of next-generation sequencing (NGS) technologies revolutionized the means by which scientists extract genetic information from biological systems, and revealed the virtually limitless insight that can be gained from the genome, transcriptome, and epigenome of various species. Following the lead of the ground-breaking Human Genome Project, several additional large-scale genome studies have materialized all over the world with NGS being the cornerstone of these investigations. Furthermore, the application of modern NGS techniques in the clinical field has led to major breakthroughs in the identification of disease-related genes in various forms of cancer and other life-threatening ailments.

A remarkable feature of NGS technologies is their high throughput nature which results in hundreds of thousands or even millions of short-read sequences. Therefore, research in the NGS field is interdisciplinary and requires integration between biological and computational knowledge and skills. In fact, the analysis of NGS data requires intensive computational power and skillful bioinformatics personnel.

However, NGS is not without its own challenges, requiring continuous development in sequencing technologies, computational infrastructure, and bioinformatics techniques to analyze the resultant raw data in order to assemble and annotate the full-length genome and transcriptome. Such developments have led to remarkable progress in efforts to enhance the performance and coverage of sequencing, and yielded a dramatic improvement in the quality of assembled sequences. Nevertheless, issues such as short-read lengths, sequencing and platform-specific errors, and large-scale memory requirements for the assembly process remain major challenges in the field.

As part of the SpringerBriefs series, this book presents a brief overview of the history, development, methods, applications, and challenges of NGS, and is divided into three parts. Part I provides an introduction to the basics of molecular biology, algorithms, and data structures required to assist readers in understanding the more technical portions of this book (Chaps. 1 and 2). Part II discusses NGS methods and the associated platforms, applications, challenges, and recent advancements

(Chaps. 3–7). Lastly, Part III provides an overview of NGS assembly stages and the related assessments and evaluations, utilized tools and remaining challenges in the field (Chaps. 8–11).

The primary audience intended for this book is newcomers to the field of sequencing with either a biological or computer science background. We provide basic introduction to both these scientific areas in relation to sequencing to allow readers to appreciate the unique amalgamation between the two that has pushed forward modern developments in the area. Furthermore, the book will also be useful for readers with a seasoned background in sequencing, as Parts II and III include a comprehensive topical review of the field including discussions on prevailing stumbling blocks in relation to technical complications, widespread availability, and the continuing need for various resources. In addition, Chap. 10 will present a unique glimpse into the recent, yet rapidly developing, field of the assessment of the next-generation sequence assembly.

California, USA   Sara El-Metwally, M.Sc.
Michigan, USA   Osama M. Ouda, Ph.D.
Toronto, Canada   Mohamed Helmy, Ph.D.

Acknowledgments

We have been fortunate to have Melanie Tucker (the former Springer editor), Meredith Clinton, and Noreen Henson as managing editors to this book. The present book started life through an initial call for book proposals sent by Melanie. Following several email communications, we met at the 13th International Conference on Systems Biology (ICSB'12, Toronto, Canada), where the proposal was pushed forward with her encouragement and enthusiasm. By the end of the project, Meredith and Noreen had also joined up to become editors of this book. Furthermore, we especially thank Dr. Suhail Asrar for language editing this whole book.

We are also grateful to the authors and developers of the methods, algorithms, and software tools in the field of next-generation sequencing who responded to our email requests. Their generosity in sharing papers and technical details was a great help to formulate the comparisons and reviews found in this book. Furthermore, we acknowledge the various community resources (web pages, wiki pages, Wikipedia pages) developed by the authors or users of methods and software tools, which represent a dynamic source of valuable practical information that is otherwise difficult to obtain. In particular, we would like to mention the NGS Field Guide by Travis Glenn, which is updated annually online.

Finally, we thank our families for their endless love and support.

Contents

Part I
Introduction to Molecular Biology and Bioinformatics Basics in Next-Generation Sequencing

Chapter 1
Basics of Molecular Biology
for Next-Generation Sequencing

Abstract Organisms can be divided into simple (or unicellular) organisms and complex (or multicellular) organisms. Both simple and complex organisms share major cellular and biological processes that are mediated through proteins and nucleic acids. Proteins are the molecules responsible for every structural or biological process achieved inside living cells or living organisms, while nucleic acids encode the necessary information required for the building and regulation of proteins. In this chapter, we present some basics of molecular biology to provide readers without a biological background with an adequate introduction to the subject. These basics would greatly aid a lay audience in understanding this and other computational biology resources and textbooks. Readers with a firm biological background may choose to skip this chapter.

1.1 Molecular Biology

Molecular Biology can be defined as the study of the molecular principles that govern and regulate biological processes. These biological processes, including the replication, transcription, and translation of genetic material, require the existence, interaction, and regulation of thousands of proteins and their corresponding genes. Thus, the focus of molecular biology starts at divulging and understanding the structure and function of these proteins/genes and continues with the study of the interactions and regulation processes between them. Additionally, the effects of their absence and mutational changes should also be understood [1, 2].

There are major differences that exist between different organisms at the molecular level, which is critical to the diversity observed between simple and complex organisms. In general, organisms are classified into two major classes according to

S. El-Metwally et al., *Next Generation Sequencing Technologies*
and Challenges in Sequence Assembly, SpringerBriefs in Systems Biology 7,
DOI 10.1007/978-1-4939-0715-1_1, © The Authors 2014

Table 1.1 Main differences between prokaryotes and eukaryotes

	Prokaryotes	Eukaryotes
Nucleus	Absent	Present
Chromosomes number	One	More than one
DNA	Circular protein-free	Liner, with chromatin
Membrane bound nucleus	Absent	Present
Telomeres	Absent (not needed)	Present
Endoplasmic reticulum	Absent	Present
Mitochondria	Absent or rare	Present
Ribosome	Small	Large
Mitosis	No	Yes
Cell wall	Chemically complex and always present	Simple and only in plants
Cell size	Small (<5 μm)	Large (>10 μm)
Unicellular/multicellular	Always unicellular	Often multicellular
Cytoskeleton	Absent	Present
Reproduction	Always asexual	Asexual or sexual
Metabolic pathways	Varity of pathways	Common set of pathways
Examples	Bacteria and archaea	Plants, animals, fungi

their cellular and genomic structures. Hence, simple organisms with a unicellular structure are called prokaryotes, while more complex organisms that are usually multicellular are called eukaryotes. The differences between prokaryotes and eukaryotes are not simply limited to the number of structure forming cells, but include several other aspects that are of great importance to the topic of this book. A major difference between them is highlighted by the structure of the genome, which is circular and protein-free with the noticeable absence of telomeres in prokaryotes. On the other hand, the eukaryotic genome may possess telomeres and proteins, which is vital for chromatin formation. These differences hold considerable influence in the processes of genome sequencing and the assembly of sequenced genomes, as will be discussed later. The major differences between prokaryotes and eukaryotes are summarized in Table 1.1 [1, 2].

Since the field of molecular biology concerns a comprehensive understanding of the structures of molecules and interactions between them, it overlaps with other fields such as biochemistry. Furthermore, with the advent of modern experimental and analytical tools such as next-generation genome sequencing (NGS) and liquid chromatography mass spectrometry (LC-MS/MS) that generate huge amounts of data due to its high-throughput nature [3], molecular biology developed the need for specialized computational and informatics tools to analyze and process this information. As a result, molecular biology has since considerably overlapped with the field of computational biology and bioinformatics [1, 2].

To develop an understanding of the details of the cellular processes at the molecular level, three particular types of molecules need to be better appreciated: deoxyribonucleic acid (DNA), ribonucleic acid (RNA), and proteins. In the next sections, we will provide a brief introduction to each of these structures.

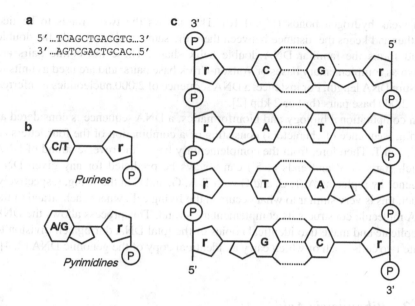

Fig. 1.1 DNA structure. (**a**) Example of complementary bases of 12 base pairs (bp). (**b**) Schematic representation of the DNA nucleotides from purines (*single ring*) and pyrimidines (*double ring*). (**c**) Structure of DNA double-stranded helix. (*r*) Deoxyribose sugar, (*P*) the phosphate group, and the *dotted lines* represent the hydrogen binds between the nucleotides of the two strands

1.1.1 Deoxyribonucleic Acid

DNA is one of the two nucleic acids that exist in living organisms and play a crucial role in cell biology. DNA is a double-stranded chain formed by the repetition of similar basic units called nucleotides. Each nucleotide consists of a sugar molecule called 2'-deoxyribose, phosphate residue, and a nitrogenous base. The sugar molecule contains five carbon atoms (arranged from 1' to 5'). The phosphate residue is important for creation of the chain through the connecting the 3' carbon atom of the sugar molecule of one nucleotide with the 5' carbon atom of the sugar molecule of the next nucleotide. Therefore, the DNA molecule has an orientation that begins at the 5' end and ends at the 3' end (Fig. 1.1a). This feature is especially important in DNA sequencing and during sequence assembly as will be discussed later. All DNA sequences available in databases, literature, or books are written from 5' to 3' unless otherwise mentioned [2, 4].

The nitrogenous bases are attached to the 1' carbon atom of the nucleotide. There are four types of bases: adenine (A), guanine (G), cytosine (C), and thymine (T). Consequently, there are four types of nucleotides described. Adenine and guanine belong to a group called purines, while cytosine and thymine belong to the pyrimidine group (Fig. 1.1b). When DNA forms the double strand, a nucleotide from the purine group is bound to a nucleotide from the pyrimidine group in the other strand. Adenine is always bound to thymine, while guanine is always bound to cytosine with weak hydrogen bonds (Fig. 1.1c). This allows the two strands to be tied together and keeps the distance between them the same, and as a result, the double strand forms the familiar DNA double helix shape. These nucleotide pairs are known as complementary bases or Watson-Crick base pairs, and are used as units to measure DNA length. For instance, a DNA sequence of 2,000 nucleotides is referred to as 2,000 base pairs (bp) or 2 kbp [2].

In computational biology and bioinformatics, a DNA sequence is considered as a string (sequence of characters) consisting of a combination of the four letters A, G, C, and T. Therefore, from the complementary bases, the reverse strand of DNA (which starts at 3′ and ends at 5′) can always be predicted for any given DNA sequence by replacing A, T, C, and G with T, A, G, and C in the string, respectively. In fact, this is very similar to what occurs in the living cell, where each strand of the DNA molecule constructs its complementary strand. This process allows the DNA to replicate and make two identical copies of the total DNA during cell division to create two cells, each of which carry an identical copy of the genomic DNA [2, 4].

1.1.2   Ribonucleic Acid

The second type of nucleic acid that exists in living cells is RNA. RNA has the same general structure and properties of DNA with certain major differences. Unlike DNA, RNA is single stranded with the sugar molecule in its nucleotides being ribose rather than 2'-deoxyribose. Furthermore, the thymine (T) base is absent and another base called uracil (U) exists instead. As a result, uracil (U) binds with adenine (A) in a similar fashion to the thymine (T) binding observed in DNA. Therefore, the RNA sequence can be predicted from the DNA sequence and vice versa through the substitution of A, U, C, and G, by T, A, G, and C, respectively. However, a major difference between DNA and RNA is that the former performs one principle function (the encoding of the genetic information of the organism) while several different types of RNA exist to accomplish a variety of tasks. It is also important to note that RNA can also exist in double strands. In some viruses, it has been observed that the genetic material is double-stranded RNA (ds-RNA) rather than DNA [2, 5]. This viral ds-RNA plays an important role in the detection of viruses by immune systems such as in humans [6].

There are three main types of RNA that play a crucial role in the protein synthesis process: the messenger-RNA (mRNA), ribosomal-RNA (rRNA), and transfer-RNA (tRNA). Furthermore, several additional forms of RNAs exist in the cell to perform critical posttranscriptional modifications and regulatory functions (Table 1.2). Here, we will briefly introduce the major types of RNA that are important to the next-generation sequencing field.

The mRNA results from DNA transcription (Fig. 1.2a), a process that creates a strand of RNA that complements a certain part of the genomic DNA (see below). This RNA is encoded such that it is actually carrying all the information needed to create the protein through the translation of the "encoded" RNA sequence into an

Table 1.2 Examples of RNAs with regulatory and posttranscriptional modification functions

RNA type	Function(s)	Organism	References
Small nuclear RNA (snRNA)[PTM]	Splicing	Eukaryotes and archaea	[20]
Y RNA[PTM]	RNA processing, DNA replication	Eukaryotes (animals)	[21]
Telomerase RNA[PTM]	Telomere synthesis	Most eukaryotes	[22]
Small nucleolar RNA (SnoRNA)[PTM]	Nucleotide modification of RNAs	Eukaryotes and archaea	[23]
Antisense RNA (aRNA)[R]	Transcriptional attenuation, mRNA degradation and stabilization	All organisms	[24]
CRISPR RNAs[R]	Resistance to bacteriophage, prevent plasmid conjugation	Bacteria	[25]
Trans-encoded base pairing sRNAs[R]	Regulation of translation and stability of target mRNAs	Bacteria	[25]
Cis-encoded base pairing sRNAs[R]	Expression regulation	Bacteria	[25]
Small interfering RNA (siRNA)[R]	Gene regulation	Eukaryotes	[26]

[R]Regulatory RNA
[PTM] Posttranscriptional modification RNA

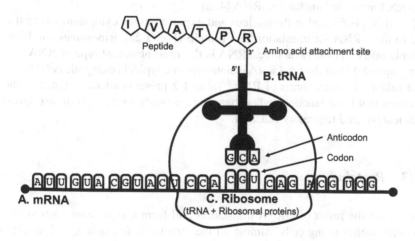

Fig. 1.2 Types of RNA and Translation process. (**a**) messenger-RNA (mRNA). (**b**) transfer-RNA (tRNA). (**c**) ribosomal-RNA (rRNA)

amino acid sequence as will be described in more detail later. In prokaryotes, the mRNA is directly translated into a protein under most circumstances, while in eukaryotes, the process is more complex due to the fact that the mRNA consists of coding regions called "exons" and noncoding regions called "introns". The removal of introns from the mRNA is crucial for the creation of mature mRNA that will be translated to a protein, a process named mRNA splicing. Therefore, mRNA splicing is a major reason for next-generation sequence assembly to be more complicated in eukaryotes in comparison to prokaryotes, where the splicing process is almost absent. Following the process of splicing in eukaryotes, the mRNA is exported to the cytoplasm to be translated. Due to the absence of the nucleus and other cellular compartments, the translation of the mRNA in prokaryotes starts during the transcription process [2, 5].

The tRNA is responsible for the transfer of amino acids to the rRNA and mRNA during the translation process. It consists of a small chain of around 80 nucleotides with a special sequence called anticodon, and has another site for amino acid attachment (Fig. 1.2b). Each tRNA is specially bound with certain amino acids via its amino acid binding site, eventually transferring this amino acid for addition to the protein that is being created. Each three nucleotide sequence in the mRNA represents one codon that corresponds to a particular amino acid. The tRNA anticodon region represents the complementary sequence of these three nucleotides. The position of the amino acid in the protein that is being translated is determined by the anticodon of the tRNA that complements the mRNA codon [2]. The mRNA codon and the tRNA anticodon regions bind with each other through hydrogen bonds, allowing the amino acids to form peptide bonds between each other and therefore, allowing the polypeptide chain (the protein being translated) to grow. tRNA bound to amino acids are termed charged tRNA or aminoacylated tRNA, while amino acid free tRNA are called uncharged tRNA [4, 5].

The rRNA is formed in the nucleus and exported to the cytoplasm where it can bind to the mRNA for translation into protein (Fig. 1.2c). Ribosomes can bind to multiple mRNA at the same time. rRNA is the most abundant type of RNA, numbering up to 80 % of the total RNA isolated from a typical eukaryotic cell [5, 7].

In relation to other forms of RNA, Table 1.2 presents alternate types of these structures that have functions other than protein synthesis e.g., posttranscriptional modifications and regulatory function.

1.1.3 Proteins

Proteins are the result of mRNA translation and form a significant portion of the structures within living cells. Almost all the structural, functional, and regulatory tasks in the cell are performed through the action of proteins. A protein is a chain of amino acids that are joined together with chemical bonds called peptide bonds or amide bonds. Each amino acid consists of a central carbon atom, a hydrogen atom, an amino group (NH_2), a carboxyl group (COOH), and a side chain which distinguishes each of the 20 naturally existing amino acids from each other. The peptide bond is formed between the carboxyl group of one amino acid and the amino group of the other, releasing a water molecule. During the translation process, protein is synthesized through the arrangement of amino acids next to each other as encoded in the genetic information, and then peptide bonds are formed between them sequentially. A short chain of amino acids is called a peptide, where the amino acids are referred to as residues. Therefore, a protein of 200 amino acids can be described as a polypeptide chain with 200 residues [2, 7].

Similar to DNA and RNA having directions (5' and 3'), a protein also possesses direction as one of its ends will always end with an amino group while the other ends with a carboxyl group. The end with the amino group is called the N-terminal while the end with the carboxyl group is called the C-terminal. The nitrogen atoms, carbon atoms, and CO- form the protein's backbone, a line that begins from the N-terminal through the C-terminal. Unlike DNA and RNA, proteins are not linear starches of sequences. In fact, a protein's sequence (amino acid order) represents the protein's primary structure. The interactions between the backbone atoms forms a "local structure" termed the protein's secondary structure. An additional layer of folding gives the protein a unique three-dimensional structure called the protein's tertiary structure. In a similar manner, yet another level of packing of the protein or with a group of different proteins is known as the protein's quaternary structure [2, 7].

1.2 The Central Dogma of Molecular Biology

The central dogma of molecular biology is a description of the information flow in biological systems. It was first introduced in the middle of the twentieth century by Francis Crick, and then published in 1970 [8]. The central dogma explains a framework of information flow from genetic material to the synthesis of proteins that perform both functional and structural roles in cells. With developments and advancements in biological research methods, analysis instruments, and imaging devices, the details of the original central dogma had been altered (e.g., the addition of the RNA splicing step). Nevertheless, its main description of the basic framework remains valid today. The central dogma states three levels of information flow, from DNA (genes) to RNA (transcripts) to amino acids (proteins) in sequential steps [8]. Here, we will describe two basic steps of the central dogma, transcription and translation, as they are crucial for understanding the subsequent chapters in this book.

1.2.1 Transcription

DNA is the main source of genetic information in organisms, with some notable exceptions where the genetic material may be composed of RNA as in the case of certain viruses [9]. In accordance with the central dogma of molecular biology, genetic information transferred from one cell to another through DNA replication process, which is the first level of information transfer. The next step is transcription, which is the process of creating a piece (sequence or stretch) of mRNA that contains the genetic information stored in corresponding DNA [8]. Transcription is an enzymatic process that is managed by RNA polymerase that sequentially attaches nucleotides to the end of the newly synthesized RNA molecule. Furthermore, the process is regulated by a group of proteins known as transcription factors that bind to specific DNA sequences and control the transcription process [10].

As mentioned in the RNA section above, in prokaryotes the mRNA is directly translated into protein, while in eukaryotes, the transcription process is more complex. The genetic structure in eukaryotic cells is more complicated in comparison to prokaryotes due to the existence of exons, entrons, and untranslated regions (UTRs). Thus, another process termed mRNA splicing follows transcription. The mRNA splicing process removes introns from the mRNA and joins the exons to create mature mRNA that is ready for translation into protein [11]. Splicing can also result in several mature mRNAs from one mRNA, resulting in several proteins from a single gene accordingly termed alternative splicing variants [12].

1.2.2 Translation

The translation process, also known as the protein synthesis process, involves "translating" the genetic code, which was transferred as nucleotides from the DNA to mRNA into a chain of amino acids (protein). The translation process requires three types of RNA: mRNA, tRNA, and rRNA. The mRNA, as explained above, carries the information required to build the target protein. The tRNA transfers the amino acids sequentially one by one following the encoding of information into the mRNA. Lastly, the rRNA is a complex of two subunits that reads the mRNA code and adds the amino acids in the same order encoded in the DNA (Fig. 1.2) [13].

Genetic information is encoded into the mRNA in triplet codons, where three nucleotides in the mRNA correspond to a specific amino acid. Typically, the reading initiates with an AUG (adenine–uracil–guanine) or imitator methionine codon and ends with a UAA, UGA, or UAG stop codon. Therefore, the rRNA reads the triplet codons and attaches the aminoacylated tRNA (tRNA with added amino acid) to the matching triplet anticodon. Subsequently, a peptide bond joins the newly added amino acid with the preceding one. As the amino acid chain grows, it starts to fold in a specific conformation that confers a three-dimensional shape to the final protein. The translation of mRNA to protein in prokaryotic cells usually occurs in the same vicinity as the transcription process due to the fact that prokaryotic cells do not possess a nucleus. In contrast, transcription takes place in the nucleus of eukaryotic cells after which the mRNA is transferred to the cytoplasm where the translation process can be achieved [1, 13].

1.3 Genetic Information Sources Targeted by Sequencing

The main target of sequencing technologies is to decode the genetic information stored in the molecules of the organisms. With modern sequencing technologies, the genetic information sources became increasingly ubiquitous, involving a myriad of molecules that lead to the development of an organism. Furthermore, special techniques have been utilized to decode the nucleotide sequences that interact/bind with other non-DNA or RNA molecules such as proteins. However, in the following paragraphs we will introduce four major types of genetic information sources that are the primary targets of available sequencing methods and platforms. In later chapters, we will also elaborate on the details of several other sources and applications.

1.3.1   The Genome

The genome represented the main target of sequencing efforts as it contained the entire genetic information of an organism. Most genomes are DNA with the exception of certain viral genomes that are RNA-based. In prokaryotes, the genome simply consists of one circular chromosome with most of its sequence represented by coding sequence which can be transcribed to RNA and then translated to proteins. In eukaryotes, the genome is far more complex, existing inside a nucleus and consisting of several pieces each of which is a separate chromosome. In most cases, the eukaryotes carry two copies of each chromosome in each cell except for the gametes in sexually reproductive organisms, which carry only a single copy. Furthermore, the genomes of eukaryotes contain noncoding regions and long intragenic stretches that are not known to encode any genetic information. Such complications bring greater challenges to whole genome sequencing (WGS) technologies and methods as well as the assembly and annotation of the sequenced genomes [1, 4].

1.3.2   The Transcriptome

The transcriptome is the entire set of RNA molecules within a single cell or population of cells. It includes the three main types of RNA (mRNA, tRNA, and rRNA) as well as the short and noncoding RNAs [1, 5]. The transcriptome represents the genes expressed at a given time (such as the time of sample collection). Therefore, it may dynamically change based on age, surroundings, media condition, and treatment of the cell/cell population. Traditionally, gene expression or the transcriptome is measured using DNA microarray techniques that allow for the measurement of a large number of genes simultaneously [14]. However, transcriptome sequencing, also known as RNA sequencing or RNA-seq, became the technology of choice for gene expression studies as its coverage is broader and allows the investigation of known and new transcripts, unlike DNA microarray techniques [15]. Similar to DNA sequence, there are two methods to assemble transcriptome sequence reads in the next-generation environment. These include the utilization of reference sequences or de novo transcriptome assembly, both of which will be discussed later on in this book.

1.3.3   The Exome

WGS identifies the sequences of all genomic DNA in the organism, including coding and noncoding sequences. In many cases, the noncoding regions of the genome are not important to a particular study. For instance, in studies targeting the identification of mutation-based diseases, the investigation of noncoding regions are less critical since 85 % of mutations exist in the coding regions (exons). Therefore, methods for sequencing the total number of exons of the genome were developed to target the whole transcribed exons (or exome) while excluding the entire population of introns. The human exome, for instance, represents 1 % of the human genome [16], which is reflective of the time and costs involved for sequencing as well as the complexity of the analysis required for the assembly and annotation of the associated reads. Therefore, several studies rely on whole exome sequencing (WES) instead of WGS to identify mutations in cancer and inherited human disorders such as Mendelian disorders [17].

1.3.4   The Metagenomes

The Metagenomes are the genomes of several organisms that coexist in a certain environment. They are mainly used during environmental studies such as sequencing and identification of organisms in an environmental sample (e.g., water or soil). Additionally, they may also be utilized in health investigations such as the study of the gut flora of humans and other organisms [18]. Typically, this field is referred to as metagenomics or environmental genomics where the study targets the sequencing and identification of all genes of all member organisms that exist in an environmental or biological sample. Since the standard sequencing procedures of model organisms normally employ cultured clones, metagenomics represents an opportunity to explore the biology and diversity of wild microorganisms in a culture-independent environment [19]. Despite the availability of several types of metagenome sequencing in the first- and next-generation methodology, the sequencing, assembly, and annotation of metagenomes remain a formidable challenge.

References

1. Alberts B, Johnson A, Lewis J, Raff M, Roberts K et al. (2007) Molecular Biology of the Cell 5th Edition. Garland Science, New York, USA
2. Setubal C, Meidanis J (1997) Introduction to Computational Molecular Biology. PWS Publishing, Pacific Grove, CA, USA
3. Helmy M, Sugiyama N, Tomita M, Ishihama Y (2010) Onco-proteogenomics: a novel approach to identify cancer-specific mutations combining proteomics and transcriptome deep sequencing. Genome Biol 11. Doi 10.1186/Gb-2010-11-S1-P17

4. Ridley M (2013) Genome: The Autobiography of a Species in 23 Chapters Harper Perennial New York, USA

5. Yarus M (2012) Life from an RNA World: The Ancestor Within. Harvard University Press Cambridge, MA, USA

6. Helmy M, Gohda J, Inoue J, Tomita M, Tsuchiya M et al. (2009) Predicting novel features of toll-like receptor 3 signaling in macrophages. PLoS One 4 (3):e4661. doi:10.1371/journal.pone.0004661

7. Meyers RA (ed) (2006) Proteins Wiley-Blackwell, Hoboken, NJ, USA

8. Crick F (1970) Central dogma of molecular biology. Nature 227 (5258):561-563

9. Patton JT (ed) (2008) Segmented Double-stranded RNA Viruses: Structure and Molecular Biology. Caister Academic Press, Poole, UK

10. Lee TI, Young RA (2000) Transcription of eukaryotic protein-coding genes. Annu Rev Genet 34:77-137. doi:10.1146/annurev.genet.34.1.77

11. Roy SW, Gilbert W (2006) The evolution of spliceosomal introns: patterns, puzzles and progress. Nat Rev Genet 7 (3):211-221. doi:nrg1807

12. Tilgner H, Knowles DG, Johnson R, Davis CA, Chakrabortty S et al. (2012) Deep sequencing of subcellular RNA fractions shows splicing to be predominantly co-transcriptional in the human genome but inefficient for lncRNAs. Genome Res 22 (9):1616-1625. doi:10.1101/gr.134445.111

13. Griffiths JFA, Wessler SR, Lewontin RC, Carroll SB (2008) Introduction to Genetic Analysis (Ninth Edition). W. H. Freeman and Company, New York, USA

14. Bowtell D, Sambrook J (2002) DNA Microarrays: A Molecular Cloning Manual. Cold Spring Harbor Lab Press, New York, USA

15. Wang Z, Gerstein M, Snyder M (2009) RNA-Seq: a revolutionary tool for transcriptomics. Nat Rev Genet 10 (1):57-63. doi:10.1038/nrg2484

16. Ng SB, Turner EH, Robertson PD, Flygare SD, Bigham AW et al. (2009) Targeted capture and massively parallel sequencing of 12 human exomes. Nature 461 (7261):272-276. doi:10.1038/nature08250

17. Kuhlenbaumer G, Hullmann J, Appenzeller S (2011) Novel genomic techniques open new avenues in the analysis of monogenic disorders. Hum Mutat 32 (2):144-151. doi:10.1002/humu.21400

18. Eisen JA (2007) Environmental shotgun sequencing: its potential and challenges for studying the hidden world of microbes. PLoS Biol 5 (3):e82. doi:1544-9173-5-3-e82 [pii]

19. Hugenholtz P, Goebel BM, Pace NR (1998) Impact of culture-independent studies on the emerging phylogenetic view of bacterial diversity. J Bacteriol 180 (18):4765-4774

20. Lui L, Lowe T (2013) Small nucleolar RNAs and RNA-guided post-transcriptional modification. Essays Biochem 54:53-77. doi:10.1042/bse0540053

21. Hall AE, Turnbull C, Dalmay T (2013) Y RNAs: recent developments. Biomolecular Concepts 4 (2):103–110. doi:10.1515/bmc-2012-0050

22. Zhou J, Ding D, Wang M, Cong YS (2014) Telomerase reverse transcriptase in the regulation of gene expression. BMB Rep 47 (2):8-14. doi:2638

23. Bachellerie JP, Cavaille J, Huttenhofer A (2002) The expanding snoRNA world. Biochimie 84 (8):775-790. doi:S0300908402014025

24. Brantl S (2002) Antisense-RNA regulation and RNA interference. Biochim Biophys Acta 1575 (1-3):15-25. doi:S0167478102002804

25. Waters LS, Storz G (2009) Regulatory RNAs in bacteria. Cell 136 (4):615-628. doi:10.1016/j.cell.2009.01.043

26. Ahmad K, Henikoff S (2002) Epigenetic consequences of nucleosome dynamics. Cell 111 (3):281-284. doi:S0092867402010814

Chapter 2
Algorithms and Data Structures in Next-Generation Sequencing

Abstract This chapter provides an overview of prevalent data structures and algorithms that are commonly utilized in bioinformatics. In particular, we place emphasis on data structures and algorithms that are employed in bioinformatic techniques during next-generation sequence assembly.

2.1 Data Structures

From a computational point of view, DNA, RNA, and proteins can be regarded as merely strings that consist of a finite set of letters (comprising of individual four letter alphabets for DNA and RNA, and a 20-letter alphabet for proteins). The efficient processing of these strings is necessary for almost all assembly and error correction techniques that are applied to bioinformatics sequences. Therefore, such techniques make extensive use of several data structures such as suffix/prefix trees, suffix arrays, graphs, hash tables, and Bloom filters. In this section, we will briefly describe these data structures and explain their applications in bioinformatics algorithms.

2.1.1 Strings

A string is a finite sequence of elements, typically characters, which are chosen from a set called an alphabet. For example, if Σ is a nonempty finite set that denotes an alphabet, then elements of Σ are called symbols or characters and any finite sequence of characters from Σ is called a string over Σ [1]. Thus, DNA sequences can be considered as strings over $\Sigma = \{A, G, C, T\}$ while RNA sequences can be regarded as strings over $\Sigma = \{A, G, C, U\}$.

The string length is a nonnegative number that indicates the number of characters in the string. A string that does not contain any characters is called an empty string (denoted by ε). The length of an empty string is 0. Certain sets of strings are of

S. El-Metwally et al., *Next Generation Sequencing Technologies*
and Challenges in Sequence Assembly, SpringerBriefs in Systems Biology 7,
DOI 10.1007/978-1-4939-0715-1_2, © The Authors 2014

special interest. For example, the set of all strings over Σ (denoted as Σ^*) is the Kleene closure of Σ whereas the set of all strings over Σ that have a specific length n is denoted as Σ^n ($\Sigma^0 = \{\varepsilon\}$ for any alphabet Σ). In the context of bioinformatics, subsequences of a fixed length n are usually called k-mers. These k-mers can be used to identify regions of interest within bioinformatics sequences. They are also very useful in the determination of sequence alignment and assembly algorithms. Consider the following case for DNA where $\Sigma = \{A, G, C, T\}$, then $\Sigma^2 = \{AA, AG, AC, AT, GA, GG, GC, GT, CA, CG, CC, CT, TA, TG, TC, TT\}$, and $\Sigma^* = \Sigma^0 \cup \Sigma^1 \cup \Sigma^2 \cup \ldots$

Defining an order in a set of strings is vital for some applications. For example, strings in suffix arrays (see below) are ordered lexicographically. Under this ordering scheme, strings are arranged based on the alphabetical order of their component characters. For example, if $\Sigma = \{A, G, C, T\}$, strings over Σ are ordered lexicographically based on the relationship $\varepsilon < A < AA < AAA < \ldots < AAAC < AAAG < A AAT < AAC \ldots$

A string of length n has n suffixes/prefixes varying in length from 1 to n. For example, the string ACGT has four prefixes (A, AC, ACG, and ACGT) and four suffixes (T, GT, CGT, ACGT).

2.1.2 Suffix/Prefix Trees

Suffix/prefix trees for a string, or a set of strings, are created by entering all the suffixes/prefixes of this string(s) into a tree structure [2]. Suffix trees in particular are very useful in solving complex string problems. A suffix trie, also called a keyword tree [3], is a special kind of suffix tree. In a suffix trie, every edge is labeled by a single character. A suffix tree is formed by concatenating all the internal nodes in the suffix trie. Figure 2.1 shows an example that illustrates the difference between the suffix trie and suffix tree for the string ACTAG. In this figure, each leaf is labeled by a number that indicates the starting position i of the corresponding suffix in the string.

In one of many beneficial applications, suffix trees can be used efficiently to search for a specific string pattern in a large collection of strings [3]. In fact, using only $O(n)$ time, a string of length n can be searched for in any collection of strings regardless of the size of the collection. For this reason, several assembly and error correction techniques for NGS data make use of suffix trees. The usage of suffix trees is limited, however, due to its large space requirement. As such, a suffix tree requires $O(n|\Sigma|\log n)$ bits to represent a string of length n over an alphabet Σ [2].

2.1.3 Suffix Arrays

Suffix arrays have similar functionality as suffix trees but with reduced space requirements. A suffix array requires $O(n \log n)$ bits only to store a string of length n over an alphabet Σ, regardless of the size of Σ [2].

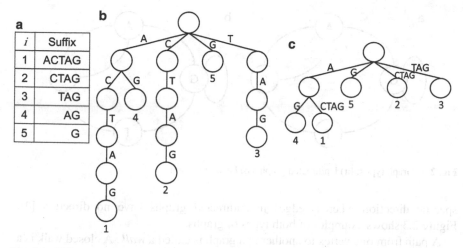

Fig. 2.1 Suffix trie versus suffix tree. (**a**) The suffixes, (**b**) suffix trie, (**c**) suffix tree for the string ACTAG

i	Suffix
1	ACTAG
2	CTAG
3	TAG
4	AG
5	G

i	SA[i]	Suffix
1	1	ACTAG
2	4	AG
3	2	CTAG
4	5	G
5	3	TAG

Fig. 2.2 Suffix array versus suffix tree. Suffix array for the string in Fig. 2.1

In a suffix array, suffixes are sorted lexicographically in increasing order. Figure 2.2 shows the suffix array (SA) of the same string in Fig. 2.1. Note the correspondence between the order of the suffixes in the suffix array and the leaves in the suffix tree.

Since the suffix array stores only the start positions of the ordered suffixes of a string of length n, each entry in the array needs $\log n$ bit of space, and since there are n entries (number of suffixes) in a suffix tree, the space requirement of a suffix array is $O(n \log n)$.

2.1.4 Graphs

A graph is a set of nodes, also called vertices, that are connected by edges. Graphs can be categorized as directed or undirected. Edges in directed graphs have a

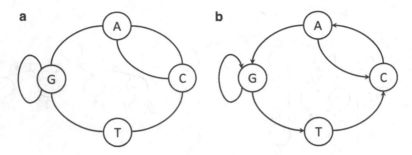

Fig. 2.3 Graph types: (**a**) Undirected graph. (**b**) Directed graph

specific direction whereas edges in undirected graphs have no direction [4]. Figure 2.3 shows examples of both types of graphs.

A path from one vertex to another in a graph is called a *walk*. A closed walk is a walk that starts and ends at the same vertex. A closed walk that has no repeated edges or vertices (other than the starting and ending vertex) is called a *cycle*, e.g., the walk A–G–G–T–C–A in Fig. 2.3b represents a cycle.

There are two types of cycles that are very useful for genome sequence assembly, namely, *Hamiltonian* cycles and *Eulerian* cycles. In a Hamiltonian cycle, every node of the graph is visited exactly once, whereas in an Eulerian cycle, every edge of the graph is visited exactly once.

Genome sequence assembly techniques which are based on Sanger sequencing make use of graphs to reconstruct long contiguous sequences from shorter reads by representing each read by a vertex and representing overlap between reads by edges that connect each pair of reads as illustrated in Fig. 2.4a. Therefore, the problem of assembling reads represented by graph vertices is reduced to finding a Hamiltonian cycle in the graph [5].

In the next-generation sequencing environment, the instruments utilized produce billions of short sequencing reads. As such, finding a Hamiltonian cycle in a graph that contains a very large number of nodes is a challenging computational problem. Therefore, the method described above for representing reads in a graph is not suitable for next-generation sequencing data.

Fortunately, discovering the location of an Eulerian cycle is much more efficient. However, this requires representing next-generation sequence reads as edges rather than vertices. To realize this, modern next-generation sequence assemblers utilize De Bruijn graphs. De Bruijn graphs were originally designed to find the shortest circular superstring that contains all possible substrings of a specific length k over a given alphabet [5]. This problem is referred to as the superstring problem. The analogy between the superstring problem and the problem of assembling billions of short sequencing reads into a single genome has attracted several researchers to De Bruijn graphs.

An important factor in the assembly of next-generation sequence short reads using De Bruijn graphs is that every distinct $k-1$ prefix or suffix of each k-mer is represented by a node. Accordingly, each pair of nodes is then connected with a

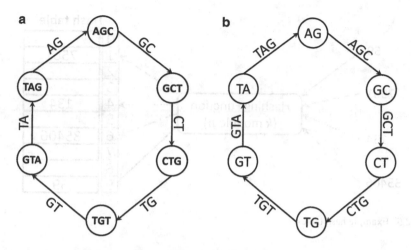

Fig. 2.4 Genome assembly using graphs. (a) Nodes and edges represent reads and overlap between reads, respectively. (b) An example of a De Bruijn graph where suffixes and prefixes of reads are represented as nodes and the corresponding k-mers as edges

directed edge if a k-mer whose prefix is one of the two nodes and whose suffix is the other node exists. For example, AGCTGTAG is a small genome sequence from which the following three short reads can be sequenced: AGCT, CTGT, and GTAG. In splitting these reads into all possible k-mers of length $k=3$, 6 different 3-mers are obtained: AGC, GCT, CTG, TGT, GTA, and TAG. For these 3-mers, we are able to list all possible $k-1$ prefixes and suffixes: AG, GC, CT, TG, GT, TA, and AG. As illustrated in Fig. 2.4b, a De Bruijn graph is constructed by representing these suffixes and prefixes as nodes and representing the corresponding k-mers (those having prefixes and suffixes as nodes in the graph) as edges. Therefore, rather than perform the computationally expensive process of finding a Hamiltonian cycle in a graph, modern assemblers prefer to identify an Eulerian cycle in De Bruijn graphs during genome reconstruction.

2.1.5 Hash Tables

The identification of repeats in a DNA sequence is critical for many applications, including the study of genome evolution and divulging the characteristics of different types of tumors. A hash table is a data structure that is commonly utilized to efficiently locate repeats within a sequence. Moreover, certain error correction techniques for next-generation sequencing, such as RACER [6], employ the use of hash tables to achieve efficient storage of k-mers. The basic idea behind hashing is simple. In a given collection of data, each data entry x is stored as a record in an array. The location of this record is computed using a hashing function $h(x)$ that assigns each data entry to a unique integer that stands for a key to that particular location in the

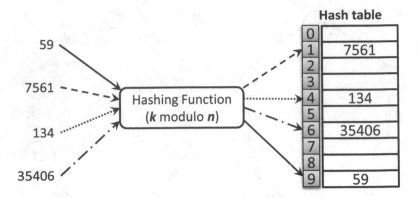

Fig. 2.5 Example hash tables

array. Accordingly, a hash table is used to preserve the set of keys that index different records in the array. Understandably, a hashing function would map similar data entries (or k-mers in the case of genome sequences) to the same index in the hash table. In other words, repeated k-mers are collected under a single slot (chained records) that is indexed using a unique key stored in a hash table [3].

Figure 2.5 shows a simple example of hashing functions. In this case, each entry (k) is stored in the hash table at position (k modulo n), where n (=10) represents the size of the hash table. A similar hashing function was used for storing k-mers in RACER [6].

2.1.6   Bloom Filters

Counting k-mers is a crucial preprocessing step for several bioinformatics applications such as genome/transcriptome sequence assembly, error correction techniques for next-generation sequence reads, and metagenomic sequencing. Several k-mer counting and abundance analysis software packages have been presented recently [7, 8]. In order to ensure rapid and memory-efficient counting of k-mers, these packages depend on memory-efficient data structures such as Bloom filters. A Bloom filter is an efficient probabilistic data structure that tells us whether an element is present or not in a data set. Bloom filters do not return false negatives. However, the efficiency of Bloom filters comes at the expense of a controlled amount of false positives. In other words, if a Bloom filter tells us that an element does not exist in a data set, we can be assured that the element is definitely not present. However, a Bloom filter may inaccurately indicate that an element is present in a particular data set when it does not.

Figure 2.6 illustrates how Bloom filters work. A Bloom filter is simply an array of bits that are initially set to zeroes as shown in Fig. 2.6a. To store an element in a Bloom filter, the element is hashed several times using multiple hashing functions

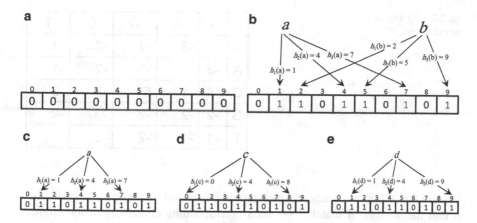

Fig. 2.6 Example of a Bloom filter of three hashing function. (**a**) A Bloom filter initialized to zeros, (**b**) two elements are inserted into the Bloom filter, (**c**) an example of true positive, (**d**) an example of true negative, and (**e**) an example of false positive

to obtain different hash values. These hash values should lie in a range between 0 and the size of the Bloom filter. Bits located at positions indicated by the resulting hash values are set to 1. Figure 2.6b–e illustrates how elements are inserted in a Bloom filter of three hash functions, and shows examples of a true positive, true negative, and false positive, respectively.

2.2 Algorithms

Bioinformatics is a field of science that studies how to relate the principles of computer science to tackle important biological questions. This can be accomplished by transforming biological queries into computational models and searching for (or developing) efficient algorithms in terms of accuracy and complexity to broach these subjects. An algorithm is a step-by-step description of a procedure of calculation, data processing, or automated reasoning [9]. In this section, we briefly discuss how a popular biological problem such as biological sequence alignment can be formulated as a computer science problem. Additionally, we will introduce algorithms and give a brief overview of a commonly utilized computer science algorithm that has been applied to a variety of biological problems, i.e., the greedy algorithm.

2.2.1 Alignment Algorithms

A typical approach to understand the functionality of a newly discovered gene is to search for close matches in a previously stored database of known genes. In order to

Fig. 2.7 Example of a
scoring matrix

	-	A	C	G	T
-		-3	-4	-2	-1
A	-3	5	-1	-2	-1
C	-4	-1	5	-3	-2
G	-2	-2	-2	-3	-2
T	-1	-1	-2	-2	5

measure how close two genes are, they should be aligned with respect to each other so that the number of matches between corresponding characters in the aligned sequences is optimized. This biological problem is analogous to the well-known computer science problem of string edit distance, which aims to measure the distance between two strings through aligning or matching them up [9]. For instance, the following alignment of the two DNA strings $x = $ CTGCG and $y = $ ACCGCT show that the number of matches between them is 3.

$$- \text{CT} - \text{GCG}$$

$$\text{AC} - \text{CGCT}$$

The alignment score is calculated based on a scoring matrix that specifies the scores of matches, mismatches, insertions, and deletions. Figure 2.7 shows an example of a scoring matrix. According to this matrix, the alignment score of the above alignment is

$$\delta(-,\text{T}) + \delta(\text{C},\text{C}) + \delta(\text{T},-) + \delta(-,\text{C}) + \delta(\text{G},\text{G}) + \delta(\text{C},\text{C}) + \delta(\text{G},\text{T}) = 7$$

There are two main classes of biological sequence alignments: *global* alignments and *local* alignments. In contrast to local alignments where only portions of sequences are aligned, the entire sequences are aligned in global alignments. Therefore, global alignments are useful for aligning closely related sequences whereas local alignments are more suitable when comparing distantly related sequences [4].

According to the number of sequences to be aligned, sequence alignment algorithms can be categorized into two categories; namely, *pairwise* alignment algorithms and *multiple* alignment algorithms. Pairwise alignment algorithms aim at finding the optimal alignment of only two sequences. On the other hand, the goal of multiple sequence alignment algorithms is to find the best alignment of three or more sequences. Figure 2.8 shows a general classification of sequence alignment algorithms.

For next-generation sequence reads, aligners take into account the application areas of next-generation sequencing technologies (e.g., metagenomics [10], cancer genomics [11], or analysis of mRNA expression [12]) as well as their unique characteristics (such as short read lengths, the large number of short reads to be mapped, and platform-dependent sequencing error rates) [13]. Therefore, such aligners have

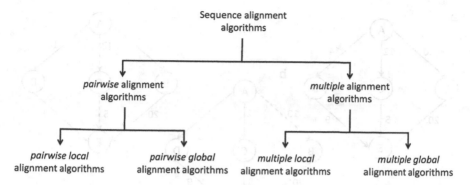

Fig. 2.8 Classification of sequence alignment algorithms

extra features compared to general alignment techniques. For example, performing alignment for short reads generated from certain next-generation sequencing technologies is computationally more intensive than the alignment of longer reads. Several short read alignment algorithms and software packages have been proposed in the last few years. Out of these aligners, Novoalign [14], SHRiMP [15], Bowtie [16], SOAP [17], Burrows-Wheeler Alignment (BWA) [18], mrFAST [19], mrs-FAST [20] are among the most popular.

2.2.2   Greedy Algorithm

Although dynamic programming can find optimal solutions for many optimization problems such as the pairwise sequence alignment issue discussed in the previous subsection, it is not always the strategy of choice for a wide range of optimization problems, especially large-scale ones, due to its high computational cost. Fortunately, the greedy algorithm provides a viable alternative strategy with reduced computational requirements. The basic idea behind this strategy is to adopt the best (optimal) choice at each possible (local) stage in the hopes that a global optimum is reached at the final stage [1]. Due to the fact that it chooses the best local solution, the greedy algorithm is also called the best first search algorithm [21]. This can be explained using the example shown in Fig. 2.9. As may be inferred from the figure, the problem here is to find the shortest path from node A to node E. Since there is no direct path from A to E, an initial decision should be made regarding the move from A to one of the three nodes B, C, or D. As shown in Fig. 2.9b, the greedy algorithm choice (dashed line) for this local step is to move to B since the distance from A to B is the shortest (best) among the three choices. It should be noted that obtaining the optimal solution using the greedy strategy is not guaranteed. This is because the greedy algorithm considers only the stage at hand and does not look ahead to the following stages. It is clear from this example that the first choice made by the algorithm does not lead to the optimal solution shown in Fig. 2.9c. Taking the next

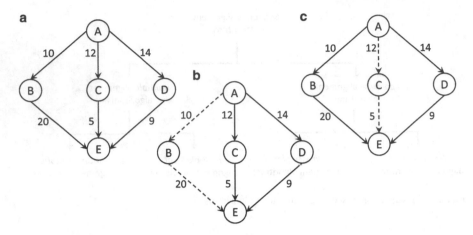

Fig. 2.9 Problem of finding the shortest path from A to E (**a**) solution using the greedy algorithm (**b**) optimal solution (**c**)

stages into consideration, one should choose node C instead of B since the distance from A to E through C is 17 (12 + 5) while the distance from A to E thorough B is 30 (10 + 20). However, the greedy algorithm often finds near-to-optimal solutions in relation to several types of optimization problems [9].

Greedy algorithms are utilized in several applications in different areas of bioinformatics. These applications include genome rearrangement and locating regulatory motifs in DNA sequences [3]. Furthermore, several recently proposed alignment and assembly algorithms for next-generation sequence data greatly benefit from the greedy algorithm. In fact, the first short read genome assemblers are based on the greedy algorithm [22–28]. The notion of applying the greedy algorithm to the next-generation sequence assembly problem is straightforward. To decide which read or contig should be added to the current one, the read (or contig) are sorted according to their overlap scores and the one with highest score is chosen [29]. However, it should be noted that next-generation sequence assemblers based on the greedy algorithms can easily get stuck at local maxima, and therefore some researchers recommend employing greedy algorithms to simple genomes only when the available computer processing power is limited.

References

1. Cormen TH, Leiserson EL, Rivest RL, Stein C (2009) Introduction to Algorithms (3rd. Ed.). MIT Press, Cambridge, MA
2. Sung WK (2010) Algorithms in Bioinformatics: A Practical Introduction. Chapman & Hall/ CRC, Boca Raton, FL, USA
3. Jones N, Pevzner P (2004) An introduction to Bioinformatics Algorithms (Computational Molecular Biology). MIT Press, Cambridge, MA, USA

4. Chacko E, Ranganathan S (eds) (2011) Chapter10: Graphs in Bioinformatics. in Algorithms in Computational Molecular Biology: Techniques, Approaches and Applications. John Wiley & Sons, Inc., Hoboken, NJ. doi:10.1002/9780470892107.ch10

5. Compeau PEC, Pevzner PA, Tesler G (2011) How to apply de Bruijn graphs to genome assembly. Nat Biotech 29 (11):987-991. doi:10.1038/nbt.2023

6. Ilie L, Molnar M (2013) RACER: Rapid and accurate correction of errors in reads. Bioinformatics. doi:btt407

7. Melsted P, Pritchard J (2011) Efficient counting of k-mers in DNA sequences using a bloom filter. BMC bioinformatics 12 (1):1-7. doi:10.1186/1471-2105-12-333

8. Zhang Q, Pell J, Canino-Koning R, Chuang Howe CA, Brown T (under review) These are not the k-mers you are looking for: efficient online k-mer counting using a probabilistic data structure. Preprint arXiv:1309:2975. In review, PloS One

9. Dasgupta S, Papadimitriou C, Vazirani U (2006) Algorithms. McGraw-Hill Science/Engineering/Math, Berkshire, UK, USA

10. Qin J, Li R, Raes J, Arumugam M, Burgdorf KS et al. (2010) A human gut microbial gene catalogue established by metagenomic sequencing. Nature 464 (7285):59-65. doi:10.1038/nature08821

11. Guffanti A, Iacono M, Pelucchi P, Kim N, Solda G et al. (2009) A transcriptional sketch of a primary human breast cancer by 454 deep sequencing. BMC Genomics 10:163. doi:10.1186/1471-2164-10-163

12. Sultan M, Schulz MH, Richard H, Magen A, Klingenhoff A et al. (2008) A global view of gene activity and alternative splicing by deep sequencing of the human transcriptome. Science 321 (5891):956-960. doi:10.1126/science.1160342

13. Li H, Homer N (2010) A survey of sequence alignment algorithms for next-generation sequencing. Brief Bioinform 11 (5):473-483. doi:10.1093/bib/bbq015

14. Novocraft Technologies (2012) Novoalign. http://novocraft.wordpress.com/2012/07/02/novoalign-v2-08-02-novoaligncs-v1-02-02-and-novosort-v1-0-released/

15. Rumble SM, Lacroute P, Dalca AV, Fiume M, Sidow A et al. (2009) SHRiMP: accurate mapping of short color-space reads. PLoS Comput Biol 5 (5):e1000386. doi:10.1371/journal.pcbi.1000386

16. Langmead B, Trapnell C, Pop M, Salzberg SL (2009) Ultrafast and memory-efficient alignment of short DNA sequences to the human genome. Genome Biol 10 (3):R25. doi:10.1186/gb-2009-10-3-r25

17. Li R, Li Y, Kristiansen K, Wang J (2008) SOAP: short oligonucleotide alignment program. Bioinformatics 24 (5):713-714. doi:10.1093/bioinformatics/btn025

18. Li H, Durbin R (2009) Fast and accurate short read alignment with Burrows-Wheeler transform. Bioinformatics 25 (14):1754-1760. doi:10.1093/bioinformatics/btp324

19. Alkan C, Kidd JM, Marques-Bonet T, Aksay G, Antonacci F et al. (2009) Personalized copy number and segmental duplication maps using next-generation sequencing. Nat Genet 41 (10):1061-1067. doi:10.1038/ng.437

20. Hach F, Hormozdiari F, Alkan C, Birol I, Eichler EE et al. (2010) mrsFAST: a cache-oblivious algorithm for short-read mapping. Nat Methods 7 (8):576-577. doi:10.1038/nmeth0810-576

21. Russell S, Norvig P (2009) Artificial Intelligence: A Modern Approach (3rd Ed.). Prentice Hall, New Jersey, USA

22. Warren RL, Sutton GG, Jones SJ, Holt RA (2007) Assembling millions of short DNA sequences using SSAKE. Bioinformatics 23 (4):500-501. doi:10.1093/bioinformatics/btl629

23. Simpson JT, Wong K, Jackman SD, Schein JE, Jones SJ et al. (2009) ABySS: a parallel assembler for short read sequence data. Genome research 19 (6):1117-1123. doi:10.1101/gr.089532.108

24. Dohm JC, Lottaz C, Borodina T, Himmelbauer H (2007) SHARCGS, a fast and highly accurate short-read assembly algorithm for de novo genomic sequencing. Genome Res 17 (11):1697-1706. doi:gr.6435207

25. Jeck WR, Reinhardt JA, Baltrus DA, Hickenbotham MT, Magrini V et al. (2007) Extending assembly of short DNA sequences to handle error. Bioinformatics 23 (21):2942-2944. doi:10.1093/bioinformatics/btm451
26. Reinhardt JA, Baltrus DA, Nishimura MT, Jeck WR, Jones CD et al. (2009) De novo assembly using low-coverage short read sequence data from the rice pathogen Pseudomonas syringae pv. oryzae. Genome research 19 (2):294-305. doi:10.1101/gr.083311.108
27. Kao WC, Chan AH, Song YS (2011) ECHO: a reference-free short-read error correction algorithm. Genome research 21 (7):1181-1192. doi:10.1101/gr.111351.110
28. Goldberg SM, Johnson J, Busam D, Feldblyum T, Ferriera S et al. (2006) A Sanger/pyrosequencing hybrid approach for the generation of high-quality draft assemblies of marine microbial genomes. Proc Natl Acad Sci U S A 103 (30):11240-11245. doi:0604351103
29. Miller JR, Koren S, Sutton G (2010) Assembly algorithms for next-generation sequencing data. Genomics 95 (6):315-327. doi:10.1016/j.ygeno.2010.03.001

Part II
Next-generation Sequencing Methods, Platforms, Applications and Challenges

Chapter 3
First- and Next-Generations Sequencing Methods

Abstract Ever since the double helix structure of DNA was first described by James Watson and Francis Crick, decoding the sequence of DNA nucleotides has been a primary focus for biologists. Therefore, methodology for DNA sequencing has undergone rapid development, particularly since the 1970s. In this chapter, we will provide an overview of the majority of sequencing methods including techniques from the first- and next-generation of DNA sequencing.

3.1 Introduction

DNA Sequencing is the precise process that determines the accurate ordering of the nucleotides in a DNA molecule. Following the sequencing of the first complete gene in 1972 [1], several different groups have worked diligently to develop approaches in order to sequence the DNA molecules, automate the sequencing process, enhance throughput, and increase commercialization, respectively. The year 1977 was a particularly landmark period as two major rapid DNA sequencing techniques were unveiled by Frederick Sanger from Cambridge (Sanger sequencing) and the team of Walter Gilbert and Allan Maxam from Harvard (Maxam–Gilbert sequencing) [2, 3]. In the ensuing years, a rapid pace followed in relation to the development of DNA sequencing applications. The first genome to be fully sequenced was the bacteriophage φX174 in 1977 by Sanger, while the first "long" genome to be fully sequenced was Epstein–Barr virus (170 Mbp) in 1984 [4].

3.2 DNA Sequencing Methods

DNA sequencing methods were non-automated techniques until the announcement of the first fully automated DNA sequencing platform by Applied Biosystems in 1987. This heralded a major boom in the field which led to the sequencing of several

S. El-Metwally et al., *Next Generation Sequencing Technologies and Challenges in Sequence Assembly*, SpringerBriefs in Systems Biology 7, DOI 10.1007/978-1-4939-0715-1_3, © The Authors 2014

model organisms and the great human genome-sequencing project [5]. In the following decade, DNA sequencing was pushed to the next level by the development of next-generation sequencing techniques. These included the establishment of DNA pyrosequencing [6], colony sequencing [7], massively parallel signature sequencing (MPSS) [8], and the parallelized version of pyrosequencing in the 1990s and early 2000s. However, the first commercialized DNA sequencing instrument from the next-generation of sequencers was the MPSS in the year 2000, followed by the parallelized version of pyrosequencing in 2004. This presented another landmark in DNA sequencing history as it speeded up analysis time and reduced the involved costs by sixfold [9]. Here, we will present the principles behind the major sequencing methods in the first- and next-generation sequencing fields.

3.2.1 First Generation Sequencing Methods

Amid the great discoveries in biology, the description of the structure of DNA by James Watson and Francis Crick is still considered among the most groundbreaking breakthroughs in biology. Undoubtedly, DNA sequencing has opened a window of virtually unlimited opportunity in relation to research and applications. In this section, we will briefly describe the Sanger and Maxam–Gilbert sequencing methods respectively, which together represent the first generation of genome-sequencing methods. Furthermore, we will introduce the concept of florescent DNA sequencing.

Sanger Sequencing

The Sanger sequencing method (also known as the chain-termination DNA sequencing method) was developed by the laboratory of Frederick Sanger (Cambridge, UK) and became the method of choice for DNA sequencing experiments for over two decades. The method was developed on the basis of synthesizing a complementary DNA strand, where a single strand of DNA molecule is sequenced through the use of chemically modified nucleotides. These chemically modified nucleotides are altered through the replacement of the hydroxyl group of its 3′ by a hydrogen atom (termed dideoxynucleotides). Since the hydroxyl group is required for the formation of the phosphodiester bond between the nucleotides, the presentation of these modified nucleotides prevents the elongation of the synthesized DNA. Accordingly, we are left with DNA fragments that end with one of the dideoxynucleotides. These fragments can be separated by size through the utilization of a gel slab. By supplying labeled dATP molecules with ^{32}P, the fragments can also be visualized and read on exposed X-ray film. While the Sanger sequencing method is slow and labor-intensive, it is still one of the most accurate DNA sequencing techniques to date. [2, 9–11].

Maxam–Gilbert Sequencing

Two years after the publication of Sanger's methods, Allan Maxam and Walter Gilbert (Harvard, USA) put forward their own sequencing method known as Maxam–Gilbert Sequencing or DNA Chemical Sequencing. The Maxam–Gilbert method requires radioactive labeling at the 5' of the DNA molecule by ^{32}P via kinase reaction through the utilization of gamma-^{32}P ATP. The purified double-stranded DNA is chemically treated to generate breaks of one or two nucleotides (G, A+G, C, C+T) and by controlling the concentration of the applied chemicals. The resulting modifications to DNA molecules can be averaged as one modification per molecule. These processes result in a series of labeled fragments that can be size-separated using electrophoresis via denaturing acrylamide gels and subsequently visualized on exposed X-ray film [3]. The capability of directly sequencing purified double-stranded DNA without the need to clone single-stranded DNA provides the Maxam–Gilbert method with an obvious advantage over the Sanger technique. However, continuous development and progressive improvements in the Sanger method made it the method of choice in most laboratories, especially in light of the chemical complexities and hazardous material associated with the Maxam–Gilbert technique.

Florescent DNA Sequencing

The two methods of DNA sequencing developed by Sanger and Maxam–Gilbert were based on radiolabeled nucleotides that could be detected by exposing the dried gel to X-ray. In the middle of the 1980s, an alternative labeling method was introduced. This new labeling method was invented by Leroy Hood (California Institute of Technology, USA) and utilized florescent-labeling instead of radiolabeling [12]. In this method, a fluorescently labeled primer replaces the radiolabeled primer of the former methods with a different flour for each of the four nucleotides. Instead of X-ray exposure, a raster scanning laser beam is used to provide the excitation required to detect the differentially labeled nucleotides. This alteration provided the advantage of eliminating multiple labor intensive processes and abolished several potential sources of error such as gel drying, X-ray film exposure, manual sequence reading from the gel and manual entry of the final sequence [13].

The florescent DNA sequencing method was implemented in the first commercialized and automated sequencing instrument (by Applied Biosystems Inc., USA). A number of laboratories migrated to this new technology due to the benefits of increased daily throughput and minimization of error sources relative to the former methods. Additionally, several laboratories utilized an automated pipetting station in conjunction with this sequencer to automate the upstream pipetting steps, further reducing labor and manual sources of error [13, 14]. The implementation of the polymerase chain reaction (PCR) [15] and the introduction of terminators (fluorescent dye-labeled dideoxynucleotides) [16] contributed to improvements in florescent DNA sequencing in the following years. PCR integration into the sequencing process provided the ability to perform cycled sequencing reactions using thermostable sequencing polymerases [13]. Furthermore, the terminators reduced the sequencing costs significantly as the four separate reactions could now be combined into a single reaction [13]. Despite these significant improvements in scalability and cost efficiency, the florescent DNA sequencing method still comprises of certain manual steps and error sources, such as manual sample loading and gel casting [13].

3.2.2   Next-Generation Sequencing Methods

For two decades, the Sanger sequencing method and its progressively improved variations were the favored techniques in most laboratories. This changed upon the announcement of the first commercialized DNA sequencing instrument by Applied Biosystems Inc., which greatly enhanced the scale of sequencing significantly [12]. The year 2005 marked the start of a new era in DNA sequencing, as developments in the field advanced rapidly in four parallel directions: (1) reducing sequencing costs, (2) speeding up the sequencing process, (3) automating the remaining manual processes, and (4) increasing the sequencing accuracy. In the following years, all four goals were remarkably achieved through the advent of next-generation sequencing (also known as massively parallel sequencing). For instance, the sequencing of the first human genome required about 10 years and hundreds of millions of dollars, a task that can now be achieved in a single day at a cost of around 5,000 dollars [17] or less with the new MiSeq X Ten sequencing platform announced by Illumina that will reduce the cost complete human genome sequencing to about 1,000 dollars [18]. Furthermore, the Food and Drug Administration (FDA) recently authorized a next-generation sequencing instrument (MiSeqDx by Illumina) for the first time, which will allow for the development of countless tests and applications in the domains of clinical and medical care [17]. In this section, we will briefly describe the major developments in next-generation DNA sequencing techniques in chronological order.

Massively Parallel Signature Sequencing

MPSS is a complicated DNA sequencing method that appeared in the late 1990s by Lynx Therapeutics (a company that later merged with Solexa). It is a bead-based method that utilizes complex steps involving adapter ligation and subsequent adapter encoding, with the sequences being read in increments of four nucleotides. The complexity of this method made it difficult to be commercialized, and the company failed to make an automated version of the sequencers that would allow the technique to be adopted by individual laboratories. Therefore, the utilization of the method was largely limited as DNA sequencing had to be accomplished at the company, with the technique being largely used to measure gene expression levels through cDNA sequencing. In addition to its complexity, MPSS suffers from several drawbacks, including the loss of specific sequences and sequencing bias. However, MPSS and its output resembled the properties of next-generation methods that appeared years later, such as the generation of hundreds of thousands of short DNA sequences. Subsequently, the acquisition of Solexa by Illumina in 2004 led to the development of a simpler sequencing approach, making MPSS largely obsolete [8].

Polony Sequencing

One of the earliest next-generation sequencing methods utilized to sequence the whole genome was the polony sequencing method developed by George M. Church (Harvard, USA). The involved processes can be broken into three main steps. Firstly, a paired-end tag library is constructed through the attachment of two tags to the 5′ and 3′ ends of the randomly sheared DNA to be sequenced. A 5′-phosphorylated and blunt-ended DNA is attached to any incomplete or damaged 5′ end, and an A is added to the 3′ end through a process called A-tailing treatment. Secondly, an emulsion PCR is used to amplify the template DNA obtained from the paired-end tag library created during the first step. Finally, ligation-based sequencing chemistry (which relies on the discriminatory capacities of ligases and polymerases) is used to sequence the amplified DNA fragments. Polony sequencing was initially used to sequence the genome of *E. coli* and demonstrated an accuracy greater than 99.9999 %. Furthermore, its cost efficiency is outstanding (being almost one-tenth of the cost associated with Sanger sequencing). Therefore, the invention of polony sequencing represented a promising technology that could be commercialized and become widely available to laboratories in the field [19].

Pyrosequencing

Pyrosequencing technology was initially invented by Mostafa Ronaghi and Pål Nyrén (The Royal Institute of Technology, Sweden) in 1996. It relies on a principle of "sequencing by synthesis", where a single-stranded DNA is taken and its complimentary strand is enzymatically synthesized [20]. Although the Sanger sequencing method also relies on the same principle, the pyrosequencing technique detects the activity of DNA polymerase (the release of pyrophosphate on nucleotide incorporation) rather than the detection of radiolabeled nucleotides in the termination sites. Later on, a paralleled version of the pyrosequencing method was developed to join the ranks of other next-generation sequencing methods by using emulsion PCR for DNA amplification. Emulsion PCR performs the amplification within water droplets inside an oil solution where each droplet contains one primer-coated bead attached to a single DNA template. To identify the newly added nucleotides in the synthesized DNA strand, the pyrosequencer uses luciferase to generate light for individual nucleotide detection [9]. Paralleled pyrosequencing technology was initially developed by 454 Life Sciences (later acquired by Roche).

Reversible Dye-Terminators Sequencing

The reversible dye-terminators sequencing technology (or Illumina Sequencing) is a florescent-labeling method that is based on the amplification of DNA molecules to form DNA clusters (local colonies of DNA on a slide) and the utilization of engineered nucleotides and enzymes [21]. The nucleotides are engineered through the addition of reversible terminate bases that are individually fluorescently labeled and attached in conjunction with a blocking group. The sequencing is accomplished in cycles, where the nucleotides compete to bind to the terminal of the newly formed DNA strand while the non-incorporated nucleotides are washed away during each cycle. A laser camera captures the florescent labels to identify the newly added nucleotide, following which the blocking group is removed to allow the start of the next cycle.

The reversible dye-terminators sequencing technology has several advantages over other next-generation sequencing methods. For instance, it utilizes only one enzyme while pyrosequencing technology requires several costly enzymes. Furthermore, it has optimal throughput allowing for maximum sequencing capacity due to the fact that the enzymatic reaction and florescent label capturing are decoupled. Therefore, the reversible dye-terminators sequencing technology is useful in sequencing difficult regions such as repetitive sequences and homopolymers, as well as the sequencing of large molecules and whole genomes [22].

Reversible dye-terminators sequencing technology was initially developed by Manteia Predictive Medicine, which was later acquired by Solexa. Solexa itself was later acquired by Illumina, which is now the primary developer of the methodology.

Sequencing by Ligation and Detection

In 2007, Applied Biosystems provided the DNA sequencing field with a new range of commercialized sequencers that implemented a technology called Sequencing by Oligonucleotide Ligation Detection (SOLiD). SOLiD is also known as the two-base encoding method since it uses a two-base color encoding schema for better differentiation between true sequence variations such as single nucleotide polymorphism (SNP) and sequencing errors. Similar to polony sequencing, SOLiD is a ligation sequencing method. However, it utilizes probes with dual base encoding (employing fluorescent labeled 8-mer probes to discern the two 3' most bases).

Before the initiation of sequencing, an amplification step is completed using emulsion PCR, following which a library/pool of DNA sequences of fixed length is created. This library contains all possible oligonucleotides sequences of that particular fixed length. Using a DNA ligase, the sequences in the library are ligated to the amplified DNA. The ligation preference favors the nucleotide in the ligation position [23]. The quality and length obtained during SOLiD sequencing are comparable with Illumina sequencing. Furthermore, this method involves each base to be read twice, which significantly reduces the SNP calling error rate and affords it an advantage in detecting variations during resequencing [24, 25]. However, a recent report revealed shortcomings in the use of ligation methods when sequencing palindromic sequences [26].

Semiconductor Sequencing

Semiconductor sequencing technology, also known as Ion Torrent semiconductor sequencing technology, is a member of the sequencing by synthesis methodologies as the sequencing takes place during the construction of the complementary strand of the single-stranded DNA molecule in question. It involves standard sequencing chemistry but implements a novel nucleotide detection approach. Instead of detecting nucleotides through radioactive or florescent labels, Ion Torrent semiconductor sequencing detects the hydrogen ions released through the DNA polymerization process [27]. Thus, no chemically modified nucleotides, optical devices, or special enzymes are required. Instead, an ion-sensitive field-effect transistor (ISFET) is needed to detect the released hydrogen ion. The ISFET acts as a hypersensitive ion sensor that measures the concentration of H^+ ions. Changes in the concentration of H^+ ions alter the current passing through the ISFET in a corresponding manner.

Certain sequencing methods have overcome the challenge of distinguishing between the four different types of nucleotides by labeling them with different labels. However, since semiconductor sequencing technology detects released H^+ during DNA polymerization, it is not possible to discern the specific type of nucleotide released based on this criterion. Instead, the template DNA is placed in a microwell and flooded with a single type of nucleotide. Thus, the detection of ions from the microwell will indicate that the newly added nucleotide matches the type of the nucleotides added to the microwell. The Ion Torrent semiconductor sequencer was released in early 2010 by Ion Torrent Systems Inc., a company that was later acquired by Life Technologies. This technology provides an inexpensive bench-top sequencing method with the limitation of relatively short reads.

References

1. Min Jou W, Haegeman G, Ysebaert M, Fiers W (1972) Nucleotide sequence of the gene coding for the bacteriophage MS2 coat protein. Nature 237 (5350):82-88
2. Sanger F, Coulson AR (1975) A rapid method for determining sequences in DNA by primed synthesis with DNA polymerase. J Mol Biol 94 (3):441-448
3. Maxam AM, Gilbert W (1977) A new method for sequencing DNA. Proc Natl Acad Sci U S A 74 (2):560-564
4. Sanger F, Air GM, Barrell BG, Brown NL, Coulson AR et al. (1977) Nucleotide sequence of bacteriophage phi X174 DNA. Nature 265 (5596):687-695
5. Adams MD, Kelley JM, Gocayne JD, Dubnick M, Polymeropoulos MH et al. (1991) Complementary DNA sequencing: expressed sequence tags and human genome project. Science 252 (5013):1651-1656
6. Ronaghi M, Karamohamed S, Pettersson B, Uhlen M, Nyren P (1996) Real-time DNA sequencing using detection of pyrophosphate release. Anal Biochem 242 (1):84-89. doi:S0003-2697(96)90432-7
7. Kawashima E, Farinelli L, Mayer P (2005) Method of nucleic acid amplification. Google Patents US20050100900 A1, https://www.google.com/patents/US20050100900

8. Brenner S, Johnson M, Bridgham J, Golda G, Lloyd DH et al. (2000) Gene expression analysis by massively parallel signature sequencing (MPSS) on microbead arrays. Nat Biotechnol 18 (6):630-634. doi:10.1038/76469

9. Schuster SC (2008) Next-generation sequencing transforms today's biology. Nat Methods 5 (1):16-18. doi:10.1038/nmeth1156

10. Tabor S, Richardson CC (1995) A single residue in DNA polymerases of the Escherichia coli DNA polymerase I family is critical for distinguishing between deoxy- and dideoxyribonucleotides. Proc Natl Acad Sci U S A 92 (14):6339-6343

11. Sanger F, Nicklen S, Coulson AR (1977) DNA sequencing with chain-terminating inhibitors. Proc Natl Acad Sci U S A 74 (12):5463-5467

12. Smith LM, Sanders JZ, Kaiser RJ, Hughes P, Dodd C et al. (1986) Fluorescence detection in automated DNA sequence analysis. Nature 321 (6071):674-679. doi:10.1038/321674a0

13. Mardis ER (2013) Next-generation sequencing platforms. Annu Rev Anal Chem (Palo Alto Calif) 6:287-303. doi:10.1146/annurev-anchem-062012-092628

14. Fulton LL, Wilson RK (1994) Variations on cycle sequencing. Biotechniques 17 (2):298-301

15. Prober JM, Trainor GL, Dam RJ, Hobbs FW, Robertson CW et al. (1987) A system for rapid DNA sequencing with fluorescent chain-terminating dideoxynucleotides. Science 238 (4825):336-341

16. Panussis DA, Cook MW, Rifkin LL, Snider JE, Strong JT et al. (1998) A pneumatic device for rapid loading of DNA sequencing gels. Genome Res 8 (5):543-548

17. Collins FS, Hamburg MA (2013) First FDA authorization for next-generation sequencer. N Engl J Med 369 (25):2369-2371. doi:10.1056/NEJMp1314561

18. Illumina Inc. (2014) HiSeq X Ten. http://www.illumina.com/systems/hiseq-x-sequencing-system.ilmn. Accessed 17-01-2014

19. Shendure J, Porreca GJ, Reppas NB, Lin X, McCutcheon JP et al. (2005) Accurate multiplex polony sequencing of an evolved bacterial genome. Science 309 (5741):1728-1732. doi:1117389

20. Margulies M, Egholm M, Altman WE, Attiya S, Bader JS et al. (2005) Genome sequencing in microfabricated high-density picolitre reactors. Nature 437 (7057):376-380. doi:10.1038/nature03959

21. Bentley DR, Balasubramanian S, Swerdlow HP, Smith GP, Milton J et al. (2008) Accurate whole human genome sequencing using reversible terminator chemistry. Nature 456 (7218):53-59. doi:10.1038/nature07517

22. Meyer M, Kircher M (2010) Illumina sequencing library preparation for highly multiplexed target capture and sequencing. Cold Spring Harb Protoc 2010 (6):pdb prot5448. doi:10.1101/pdb.prot5448

23. Valouev A, Ichikawa J, Tonthat T, Stuart J, Ranade S et al. (2008) A high-resolution, nucleosome position map of C. elegans reveals a lack of universal sequence-dictated positioning. Genome Res 18 (7):1051-1063. doi:10.1101/gr.076463.108

24. Rieber N, Zapatka M, Lasitschka B, Jones D, Northcott P et al. (2013) Coverage bias and sensitivity of variant calling for four whole-genome sequencing technologies. PLoS One 8 (6):e66621. doi:10.1371/journal.pone.0066621

25. Liu L, Li Y, Li S, Hu N, He Y et al. (2012) Comparison of next-generation sequencing systems. J Biomed Biotechnol 2012:251364. doi:10.1155/2012/251364

26. Huang YF, Chen SC, Chiang YS, Chen TH, Chiu KP (2012) Palindromic sequence impedes sequencing-by-ligation mechanism. BMC Syst Biol 6 Suppl 2:S10. doi:10.1186/1752-0509-6-S2-S10

27. Rusk N (2011) Torrents of sequence. Nature Methods 8 (1):44-44. doi:10.1038/Nmeth.F.330

Chapter 4
Next-Generation Sequencing Platforms

Abstract The next-generation sequencing methods that we described in Chap. 3 were implemented into several commercial instruments in order to make the sequencing technologies available to individual laboratories. In this chapter, we will overview the major DNA sequencing platforms and present several comparisons for their advantages, disadvantages, sequencing costs, and other aspects.

4.1 Introduction

In Chap. 3, we described several next-generation sequencing methods. These methods have been implemented into several commercialized sequencing platforms that allowed DNA sequencing to become a laboratory-based task, contrasting with previous models where the sequencing required specialized laboratories, centers, or companies. Several reports and studies have reviewed, compared, and benchmarked the various sequencing platforms. These studies evaluated the chemistry involved, read length, accuracy, coverage, machine price, sequencing costs, and other factors for comparisons [1–3]. Other studies assessed bioinformatics complications during data analysis and sequence assembly as measures for the usefulness of each platform [4]. Here, we will briefly introduce commonly utilized sequencing platforms as well as more recently available systems. In subsequent chapters, the bioinformatics and data analysis of these platforms will be also discussed in detail.

4.2 Roche 454 Systems

The Roche 454 systems utilize pyrosequencing technology, which detects the released pyrophosphate during nucleotide incorporation (see Chap. 3). The first commercially successful next-generation sequencer was the pyrosequencing platform that was developed by 454 Life Sciences in 2005 (454 Life Sciences was later

S. El-Metwally et al., *Next Generation Sequencing Technologies*
and Challenges in Sequence Assembly, SpringerBriefs in Systems Biology 7,
DOI 10.1007/978-1-4939-0715-1_4, © The Authors 2014

acquired by Roche). This system had a read length of 100–150 bp and output reads that exceeded 200,000 reads and 20 MB/run [1, 5]. The next system launched in 2008 (454 GS FLX Titanium) increased the read length to 700 bp and the output to 0.7 GB with 99.9 % accuracy, taking less than 1 day per analysis. In the following year, Roche launched a bench-top sequencing system called GS Junior with simplified library preparation and elevated output up to 14 GB/run [6]. Recently, Roche launched its latest system, the GS FLX+ which featured a read length up to 1,000 bp and ~1,000,000 shotgun reads/run [7]. In summary, the main advantages of Roche systems are the increased read lengths and high speed while the main disadvantages are the relatively high error rates and expensive costs of the reagents [1]. Other characteristics of Roche 454 systems are listed in Tables 4.1, 4.2, 4.3, and 4.4.

Table 4.1 Comparison of major next-generation sequencers advantages and disadvantages[a]

Instrument	Primary advantages	Primary disadvantages
ABI 3730xl (capillary)	Low cost for very small studies	Very high cost for large amounts of data
454 FLX Titanium	Long read length	High cost per MB
454 FLX+	Double the maximum read length of Titanium	High capital cost; high cost per MB; reagent issues; upgrade issues
454 GS Jr. Titanium	Long read length; low capital cost; low cost per experiment	High cost per MB
Illumina GAIIx	Lower capital cost than HiSeqs	Slightly higher cost per MB than HiSeq; not as scalable in the future
Illumina HiSeq 1000	Lower instrument cost than HiSeq 2000; same number of reads/lane and cost/lane as HiSeq 2000; field upgradable to HiSeq 2000; future scalability	Not as flexible as HiSeq 2000 due to having only one flow cell
Illumina HiSeq 2000	Same as HiSeq 1000, but runs two flow cells simultaneously; most reads, GB per day and GB per run, lowest cost per MB of all platforms[a]	High capital cost; high computation needs
Illumina HiSeq 2500	Same as HiSeq 2000, but can also run two 2 lane miniFlowCells to achieve much faster run times and longer read lengths	miniFlowCell will likely have a higher cost per read than standard HiSeq Flow Cell; can't run miniFlowCell and standard Flow Cell at the same time
Illumina HiScanSQ	Versatile instrument for full catalog of Illumina arrays and sequencing; scalable in future	Higher cost/MB than HiSeq for large amounts of data
Illumina MiSeq	Moderate cost instrument and runs; low cost per MB for a small platform; fastest Illumina run times and longest Illumina read lengths	Relatively few reads and higher cost per MB compared to HiSeq
Ion Torrent— 314 chip	Low cost per sample for small studies; short time needed on instrument; suitable for microbial sequencing and targeted sequencing	High cost per MB; sample prep takes longer than time on the instrument; far fewer reads and slightly shorter total read length than MiSeq

(continued)

Table 4.1 (continued)

Instrument	Primary advantages	Primary disadvantages
Ion Torrent— 316 chip	Same as 314, upgraded due to higher density chip; lower cost per read and MB allows more applications	Similar to 314, but more reads
Ion Torrent— 318 chip	Same as 316, upgraded due to higher density chip; lower cost per read and MB allows more applications	Similar to 316, but more reads
Ion Torrent— PGM	Low cost instrument upgraded through disposable chips (the chip is the machine); very simple machine with few moving parts; clear trajectory to improved performance	Higher error rate than Illumina; more hands-on time and fewer reads at higher cost per MB relative to MiSeq
Ion Torrent— Proton	Moderately low-cost instrument for high throughput applications; similar cost to MiSeq, but PII and PIII chips will give more reads than MiSeq	Higher error rate than Illumina; more hands-on time and shorter reads than MiSeq or HiSeq 2500 Rapid Run
Ion Torrent— Proton-I chip	Similar to PGM chips, but with many more sensors (wells); more reads than MiSeq at similar cost/MB; single contiguous read similar in length to HiSeq total length	Higher cost/MB than HiSeq; shorter reads than MiSeq; higher error rate than Illumina; more analysis tools needed
Ion Torrent— Proton-II chip	Same as Proton-I chip, but with more sensors (wells); similar or possibly lower cost/MB than HiSeq	Same as Proton I but more reads and lower cost per MB
Ion Torrent— Proton-III chip	Same as Proton-II chip, but with more sensors (wells); similar or possibly lower cost/MB than HiSeq	Same as Proton II but more reads and lower cost per MB
SOLiD—5500xl	Each lane of Flow-Chip can be run independently; high accuracy; output in bases (not color-space); ability to rescue failed sequencing cycles; 96 validated barcodes per lane; throughput of 20–30 GB/day	Longevity of platform; relatively short reads; more gaps in assemblies than Illumina data; less even data distribution than Illumina; high capital cost

[a]The data is derived from 2013 NGS Field Guide update [3, 18]

4.3   AB SOLiD Systems

SOLiD sequencing systems are based on Sequencing by Ligation and Detection technology (see Chap. 3). In 2007, Applied Biosystems Inc. (ABI) acquired SOLiD and launched a sequencing system with a read length of 35 bp and output of 3 GB/run with 99.85 % accuracy [1]. In the following 3 years, ABI released five upgrades to the system before the release of SOLiD 5500xl in 2010. The SOLiD 5500xl came with outstanding features in all respects. The read length was extended to 85 bp and the output increased to 30 GB/run with 99.99 % accuracy, though it took about 1 week to complete a run. Recently, ABI unveiled the 5500 W Series Genetic Analyzers that demonstrate a significant reduction in sequencing costs (~50 %),

Table 4.2 Comparison of major next-generation sequencers run time, read length, and output data[a]

Instrument	Run time	Millions of reads/run	Read length	Yield MB/run	Data file sizes (GB)
ABI 3730xl (capillary)	2 h	9.6×10^{-5}	650	0.06	0.03
454 FLX Titanium	10 h	1	400	400	20 images, 4 sff
454 FLX+	20 h	1	650	650	40 images, 8 sff
454 GS Jr. Titanium	10 h	0.1	400	50	<3 images, <1 sff
Illumina GAIIx	14 days	300	150+150	96,000	600
Illumina HiSeq 1000	8.5 days	≤1,500	100+100	≤300,000	≤300
Illumina HiSeq 2000	11.5 days	≤3,000	100+100	≤600,000	≤600
Illumina HiSeq 2500—rapid	40 h	≤600	150+150	≤180,000	[big]
Illumina iScanSQ	8.5 days	700	100+100	140,000	50
Illumina MiSeq—version 1	26 h	4	150+150	1,200	1
Illumina MiSeq—version 2	39 h	15	250+250	7,500	1
Ion Torrent—"314" chip	4 h	0.1	400	40	0.1 sff, 0.2 fastq
Ion Torrent—"316" chip	4 h	1.6	400	400	5 sff, 1 fastq
Ion Torrent—"318" chip	7 h	4	400	1,500	10 sff, 2.5 fastq
Ion Torrent—Proton I	≤4 h	70	≤200	10,000	120 sff, 30 fastq
Ion Torrent—Proton II	[>4 h]	[250]	[≤200]	[50,000]	[big]
Ion Torrent—Proton III	[>4 h]	[500]	[≤200]	[100,000]	[big]
SOLiD—5500xl	8 days	>1,410	75+35	155,100	148

[a]The data is derived from 2013 NGS Field Guide update [3, 18]

simplified workflow, and increased throughput, though their read length remains short (max. 75 bp) [8]. To conclude, the main advantages of SOLiD systems are the reasonable cost and enhanced accuracy while the main disadvantages are the shorter read lengths and the requirement of advanced computational resources as well as skilled bioinformatics personnel to analyze the 4 Tb of data that are generated from each run [1]. In general, SOLiD systems are suitable for whole genome and transcriptome studies. Other characteristics of Roche 454 systems are listed in Tables 4.1, 4.2, 4.3, and 4.4.

4.4   Illumina GA/HiSeq Systems

Illumina systems are based on Reversible Dye-Terminators Sequencing, which is a sequencing-by-syntheses approach (see Chap. 3). Solexa initially developed and also launched the first sequencer implementing this technology, the Genome Analyzed (GA), before being acquired by Illumina in 2007. Although the first GA only had an output of 1 GB/run, the release of the GAIIx series after a period of 3 years provided a drastic improvement with 85 GB/run and a read length of 150 bp [9]. Illumina launched the Illumina HiSeq 2000 system in 2010 with an initial output of 200 GB/run. The output was soon increased to 600 GB/run, with each run taking about a week [1]. The most recent Illumina system is the HiSeq 2500/1500,

Table 4.3 Comparison of major next-generation sequencers purchase and operation costs[a]

Instrument	Reagent cost/run	Reagent cost/MB	Minimum unit cost (% run)	Purchase cost	Additional instruments	Service contract	Computational resources
ABI 3730xl (capillary)	$144	$2,308	$6 (1 %)	$376	–	$19.80	Desktop
454 FLX Titanium	$6,200	$12	$2,000 (12 %)	$29.50	–	–	$5 (desktop)
454 FLX+	$6,200	$7	$2,000 (12 %)	$450	$30	$50	$5 (desktop)
454 GS Jr. Titanium	$1,100	$22	$1,500 (100 %)	$108	$16	$12.60	$5 (desktop)
Illumina GAIIx	$17,575	$0.19	$2,500 (14 %)	$250	$100	$44.50	$222 cluster
Illumina HiSeq 1000	$10,220	$0.04	$2,600 (12 %)	$560	$55	$62	$222 cluster
Illumina HiSeq 2000	$23,470	≥$0.04	$2,400 (6 %)	$690	$55	$75.90	$222 cluster
Illumina HiSeq 2500—rapid	$6,145	$0.05	NA (50 %)	$690	$55	$75.90	$222 cluster
Illumina iScanSQ	$12,750	$0.09	$2,500 (14 %)	$405	$55	$41.50	$222 cluster
Illumina MiSeq—version 1	$1,040	$0.70	~$1,400 (100 %)	$125	–	$12.50	Desktop/cloud
Illumina MiSeq—version 2	$1,070	$0.14	$1,400 (100 %)	$0	–	–	Desktop/cloud
Ion Torrent—"314" chip	$539	$5	~$750 (100 %)	$49	$18/32	$7.5/9.9	$16.5 (desktop)
Ion Torrent—"316" chip	$739	$1.20	~$1,000 (100 %)	$49	$18/32	$7.5/9.9	$16.5 (desktop)
Ion Torrent—"318" chip	$939	$0.60	~$1,200 (100 %)	$49	$18/32	$7.5/9.9	$16.5 (desktop)
Ion Torrent—Proton I	$1,050	$0.09	–	224	$19.5/3.3–8.5	22.4	($75) cluster
Ion Torrent—Proton II	[$1,000]	[$0.02]	–	224	$19.5/3.3–8.5	22.4	($75) cluster
Ion Torrent—Proton III	[$1,000]	[$0.01]	–	224	$19.5/3.3–8.5	22.4	($75) cluster
SOLiD—5500xl	$10,503	< $0.07	$2,000 (12 %)	$251	$54	$44.40	$35 cluster

[a]The data is derived from 2013 NGS Field Guide update [3], some values are from 2011's guide [18]. All costs are in thousands of US dollars

Table 4.4 Comparison of major next-generation sequencers errors and error rates[a]

Instrument	Data file sizes (GB)	Primary errors	Single-pass error rate (%)	Final error rate (%)
ABI 3730xl (capillary)	0.03	Substitution	0.1–1	0.1–1
454 FLX Titanium	20 images, 4 sff	Indel	1	1
454 FLX+	40 images, 8 sff	Indel	1	1
454 GS Jr. Titanium	<3 images, <1 sff	Indel	1	1
Illumina GAIIx	600	Substitution	~0.1	~0.1
Illumina HiSeq 1000	≤300	Substitution	~0.1	~0.1
Illumina HiSeq 2000	≤600	Substitution	~0.1	~0.1
Illumina HiSeq 2500—rapid	[big]	Substitution	~0.1	~0.1
Illumina iScanSQ	50	Substitution	~0.1	~0.1
Illumina MiSeq—version 1	1e	Substitution	~0.1	~0.1
Illumina MiSeq—version 2	1e	Substitution	~0.1	~0.1
Ion Torrent—"314" chip	0.1 sff, 0.2 fastq	Indel	~1	~1
Ion Torrent—"316" chip	5 sff, 1 fastq	Indel	~1	~1
Ion Torrent—"318" chip	10 sff, 2.5 fastq	Indel	~1	~1
Ion Torrent—Proton I	120 sff, 30 fastq	Indel	~1	~1
Ion Torrent—Proton II	[big]	Indel	~1	~1
Ion Torrent—Proton III	[big]	Indel	~1	~1
SOLiD—5500xl	148	A-T bias	~5	≤0.1

[a] The data is derived from 2013 NGS Field Guide update [3, 18]

which provides a similar output in 2–11 days with up to six billion reads [10]. Moreover, Illumina announced the release of the new HiSeq X Ten sequencing system in January 2014 which is specially adapted for analyzing large sample populations. The new system consists of ten ultra-high throughput sequencers that work in parallel. Interestingly, Illumina claims that the HiSeq X Ten system is the first sequencing methodology to deliver full coverage human genomes at a cost of $1,000 or less [11]. To summarize, the main advantages in Illumina systems are the high throughput and inexpensive costs involved while the main disadvantages are the short read length and advanced computational resources required to store and analyze the huge amount of data generated from each run [1]. Illumina systems are very suitable for whole genome, transcriptome, and personal genome applications. Other characteristics of Illumina systems are listed in Tables 4.1, 4.2, 4.3, and 4.4.

4.5   Compact Systems

Roche, SOLiD, and Illumina provide the major sequencing platforms being used for genome and other sequencing applications in the next-generation field. However, the continuous demand for cheaper and single-laboratory-based equipment led vendors to develop compact versions of the above sequencers or other types of sequencers with novel technologies. These compact sequencers share the same features as the full size versions except that they are smaller in size and lower in throughput, therefore being cheaper in the price. Their use is mainly targeted for personal genomics and clinical applications.

4.5.1 Illumina MiSeq

MiSeq is a compact sequencing system developed by Illumina. It implements the sequencing by synthesis technology and resembles the Illumina HiSeq system. However, MiSeq combines the processes of cluster generation, sequencing by synthesis, and data analysis into a single machine. Thus, it reduces the sequencer size to a bench-top machine and decreases the analysis time to as little as 4 h [1, 2]. Apart from its size, price, and speed, MiSeq comes with several other outstanding features. For instance, the read length of MiSeq is 300 bp, which significantly reduces the data analysis effort [10]. Thus, MiSeq is suitable for several novel applications such as clinical applications, small genomes sequencing, clone checking, and ChIP-seq [1]. Furthermore, the Illumina MiSeqDx recently became the first sequencer to be granted marketing authorization from the Food and Drug Administration (FDA) [12].

4.5.2 Ion Torrent PGM

Life Technologies released the Ion Personal Genome Machine (PGM) in 2010. The PGM implements semiconductor sequencing technology that does not require fluorescence and camera scanning, resulting in an outstandingly fast machine (2 h/ analysis). Furthermore, the sample preparation is accomplished in parallel for eight samples and takes only 6 h. The PGM read length is 200 bp on average and can be up to 400 bp, which is another major advantage [1, 13]. The PGM is an ultra fast sequencer but with limited throughput (up to 2 GB per run), and is typically used for applications such as the identification of microbial pathogens. During the 2011 outbreak of exceptionally virulent *Escherichia coli* that centered in Germany, the PGM was used in conjunction with HiSeq for whole genome sequencing to identify the type of *E. coli* involved. This information was particularly useful in helping scientists understand the associated antibiotic resistance [14–16].

4.5.3 The Open-Source Sequencing System

Complete Genomics, a wholly owned subsidiary of the Beijing Genome Institute (BGI)-Shenzhen, has its own sequencer named the Polonator G.007. The Polonator is based on the polony sequencing sequencer and represents an inexpensive, high performance, and high throughput machine with open-source software and protocols [1].

According to Complete Genomics, the accuracy of the system is 99.999 %, which is comparable to the accuracy of the HiSeq system [1]. Therefore, the Polonator would be suitable for the identification of SNPs and indels. The high level of accuracy claimed by the makers was recently supported by the discovery of a point mutation responsible for Prader-Willi Syndrome (a rare genetic disorder). In this case, the authors obtained the whole genome sequence of two healthy parents and an affected son through the utilization of sequencing technology provided by Complete Genomics [17].

References

1. Liu L, Li Y, Li S, Hu N, He Y et al. (2012) Comparison of next-generation sequencing systems. Journal of biomedicine & biotechnology 2012:251364. doi:10.1155/2012/251364
2. Quail MA, Smith M, Coupland P, Otto TD, Harris SR et al. (2012) A tale of three next generation sequencing platforms: comparison of Ion Torrent, Pacific Biosciences and Illumina MiSeq sequencers. BMC Genomics 13:341. doi:10.1186/1471-2164-13-341
3. Glenn TC (2011) Field guide to next-generation DNA sequencers. Mol Ecol Resour 11 (5):759-769. doi:10.1111/j.1755-0998.2011.03024.x
4. El-Metwally S, Hamza T, Zakaria M, Helmy M (2013) Next-generation sequence assembly: four stages of data processing and computational challenges. PLoS Comput Biol 9 (12):e1003345. doi:10.1371/journal.pcbi.1003345
5. Mardis ER (2008) The impact of next-generation sequencing technology on genetics. Trends Genet 24 (3):133-141. doi:10.1016/j.tig.2007.12.007
6. Huse SM, Huber JA, Morrison HG, Sogin ML, Welch DM (2007) Accuracy and quality of massively parallel DNA pyrosequencing. Genome Biol 8 (7):R143. doi:gb-2007-8-7-r143
7. 454.com GS FLX+ Systems http://454.com/products/gs-flx-system/index.asp. Accessed 10-01-2014
8. Life Technologies Inc. (2012) 5500 W SerieS Genetic Analyzers. http://tools.lifetechnologies.com/content/sfs/brochures/5500-w-series-spec-sheet.pdf Accessed 10-01-2014
9. Illumina Inc. (2011) Genome Analyzer IIX. http://res.illumina.com/documents/products/datasheets/datasheet_genome_analyzeriix.pdf. Accessed 10-01-2014
10. Illumina Inc. (2014) Illumina Sequencing Systems. http://www.illumina.com/systems/sequencing.ilmn. Accessed 10-01-2014
11. Illumina Inc. (2014) HiSeq X Ten. http://www.illumina.com/systems/hiseq-x-sequencing-system.ilmn. Accessed 17-01-2014
12. Collins FS, Hamburg MA (2013) First FDA authorization for next-generation sequencer. N Engl J Med 369 (25):2369-2371. doi:10.1056/NEJMp1314561
13. Life Technologies Inc. (2014) Ion Personal Genome Machine. https://www.lifetechnologies.com/order/catalog/product/4462921. Accessed 10-01-2014
14. Mellmann A, Harmsen D, Cummings CA, Zentz EB, Leopold SR et al. (2011) Prospective genomic characterization of the German enterohemorrhagic Escherichia coli O104:H4 outbreak by rapid next generation sequencing technology. PLoS One 6 (7):e22751. doi:10.1371/journal.pone.0022751
15. Rohde H, Qin J, Cui Y, Li D, Loman NJ et al. (2011) Open-source genomic analysis of Shiga-toxin-producing E. coli O104:H4. N Engl J Med 365 (8):718-724. doi:10.1056/NEJMoa1107643
16. Chin CS, Sorenson J, Harris JB, Robins WP, Charles RC et al. (2011) The origin of the Haitian cholera outbreak strain. N Engl J Med 364 (1):33-42. doi:10.1056/NEJMoa1012928
17. Schaaf CP, Gonzalez-Garay ML, Xia F, Potocki L, Gripp KW et al. (2013) Truncating mutations of MAGEL2 cause Prader-Willi phenotypes and autism. Nat Genet 45 (11):1405-1408. doi:10.1038/ng.2776
18. Glenn TC (2013) Field guide to next-generation DNA sequencers-Update. http://www.molecularecologist.com/next-gen-fieldguide-2013/. Accessed 10-01-2014

Chapter 5
Challenges in the Next-Generation Sequencing Field

Abstract The next-generation sequencing field has developed at such a rapid pace that the achievements of today exceed the challenges and limitations of the last few years. We have previously described how next-generation sequencing has advanced DNA sequencing from low throughput to high throughput and minimized labor-intensive practices to more automated processes. In this chapter, we will compare next-generation sequencing with the traditional Sanger method to demonstrate the remarkable recent developments in relation to improved speed, throughput, and accuracy. Lastly, we will provide an overview of lingering challenges that are still faced by next-generation sequencing.

5.1 Sanger Sequencing Versus Next-Generation Sequencing

Descriptions of both Sanger sequencing and the major next-generation sequencing technologies have been provided in the previous chapters. However, we would like to highlight certain differences between the two technologies in terms of both properties and applications (Table 5.1). The first major difference in relation to properties is that Sanger technology is based on cloning of the target molecule and the sequencing of smaller subclones of the original clone. The subclones are then joined together to construct the original clone, following which several clones are linked together using their overlapping ends to form a chromosome. In the next-generation technologies, rather than cloning, the target molecule is broken down into short fragments and DNA adapters are attached to the ends to construct a library of short DNA fragments that will be amplified ahead of the sequencing process [1].

Another principal difference between the two technologies is the read length of the resultant sequences. Since the Sanger method is clone-based, the resulting sequences are relatively long (800–1,000 bp), while the next-generation sequencing technologies vary from 75 to 700 bp [2]. This difference has a major impact in the

S. El-Metwally et al., *Next Generation Sequencing Technologies
and Challenges in Sequence Assembly*, SpringerBriefs in Systems Biology 7,
DOI 10.1007/978-1-4939-0715-1_5, © The Authors 2014

Table 5.1 Features of Sanger and next-generation sequencing technologies

	Sanger sequencing	Next-generation sequencing
Sample	Clones, PCR	DNA libraries
DNA cloning/amplification	Cloning	Amplification and tagging
Throughput	Low	High (massively parallel)
DNA synthesis and newly added nucleotide detection	Distinct processes	Simultaneously
Coverage	Low coverage depth	High coverage depth
Sequencing accuracy	Highly accurate	Relatively lower accuracy
Sample-read rate	One sample one read	One sample up to millions reads
Cost efficiency	Expensive	Efficient
Sequencing repetitive regions	More effective	Less effective
Read length	Longer	Shorter
Genome assembly in the absence of reference	Easier	Harder

data analysis of the resultant sequence, especially in light of the exceptional high throughput of next-generation sequencing technologies. We will further discuss the computational challenges involved in the subsequent chapters.

5.2 Efficiency of Sanger Sequencing

Following the introduction of next-generation sequencing, Sanger sequencing started to lose its once dominant foothold in the sequencing field. However, the utilization of the Sanger method still retains certain strengths [1]. At its inception, Sanger technology was originally intended for the accurate sequencing of long DNA molecules rather than whole genome sequencing. To date, the Sanger technique is still widely utilized as a cost-efficient methodology for sequencing constructs, PCR products, and individual genes. Furthermore, it remains one of the few technologies that offer accurate confirmation of mutations and findings during certain clinical applications, where it continues to hold an edge over next-generation methodologies.

One of the recent cases where Sanger sequencing demonstrated outstanding efficiency and accurate sequencing capabilities was during the late phase of the self-replicating synthetic bacterial cell project (J. Craig Venter Institute, USA) [3]. In this project, scientists computationally designed the genome of a simple bacterial cell and then synthesized the whole genome in pieces. These pieces were then joined together in several phases through the utilization of E. coli and yeast cells [4]. However, in the final stages of the project, the combined genome failed to boot-up after it had been transferred to the new host cell (a genome-free bacterial cell). It took a period of 2 years to attribute the causal factor behind this failure as an error during the genome synthesis process, where a single nucleotide was found to be missing. In this case, the Sanger method outdid the next-generation sequencers in uncovering this minute difference between the synthesized sequence and the design template [4].

5.3 Challenges in Next-Generation Sequencing

The accelerated pace of technological development is greatly exemplified by comparing the present day requirements for human whole genome sequencing with the groundbreaking Human Genome Project that was initiated over two decades ago. The associated spending costs of the Human Genome Project were an estimated 100,000 million dollars, taking over a decade to complete. In contrast, modern methodologies may reduce expenses to around 5,000 dollars, with a required time period of 14 days [1] or less with the recently announced systems by Illumina [5, 6]. Furthermore, newer compact sequencers (such as the PGM from Ion Torrent and MiSeq from Illumina) combine template preparation, sequencing, imaging, and data analysis into a single bench-top machine, significantly reducing required manpower and related costs. Therefore, the next-generation sequencing field is achieving rapid and remarkable progress to overcome previous limitations associated with technologies, methods, and protocols. However, three important challenges have yet to be overcome, i.e., costs, the read length, and the error rates and types.

5.3.1 Sequencing Cost

Despite dramatic decreases in actual sequencing expenses in recent years, the initial costs of establishing sequencing facilities remain too high for such indispensable technology. The comparisons we showed in Chap. 4 (Table 4.3) illustrate the cost of the major available sequencing platforms. This can range from 0.7 million dollars to around 100,000 dollars for the least costly models, therefore putting next-generation sequencing out of the reach of most laboratories and hospitals. With relation to developing countries, the Human Genetics Programme (HGP) at the World Health Organization (WHO) published a report on the potential health impact of DNA sequencing technologies in these nations [7]. The report revealed the potential positive impact of the technology but identified cost as a limiting factor in relation to its distribution and application. Another recent WHO report on the spread of the poliovirus in the Horn of Africa revealed the indispensability of DNA sequencing in the confirmation of laboratory results and ascertaining the origin of isolated viruses [8]. Taking current expenses into consideration, the establishment of DNA sequencing facilities in developing countries for either health applications or research remains improbable.

5.3.2 Read Length

Despite the numerous advantages of the next-generation sequencing technologies, limitations in the read length remains its major technical drawback [2]. Recent next-generation sequencers have demonstrated a range between 75 and 900 bp, though the average length is between 100 and 400 bp (Chap. 4, Table 4.2) [9, 10].

In this case, shorter read-lengths are known to require more difficult and complicated analyses. As will be discussed in more detail later, short reads are assembled together using overlapped ends to create longer stretches of DNA. Accordingly, the DNA stretches are attached to each other to elongate them further until the construction of an entire genome (for prokaryotes), a whole chromosome (for eukaryotes), or full DNA or RNA molecules occurs. In contrast, longer read-lengths greatly simplify the assembly process due to two major reasons. Firstly, short reads comprise of shorter overlapping ends, which makes the accurate determination of the preceding and following reads difficult. Secondly, longer reads require less rounds of the overall assembly process. For instance, a DNA molecule of 100,000 bp will have ~2,000 reads of length 50 bp or ~110 reads of length 900 bp. Therefore, the analysis and assembly efforts required increases by several folds for short read-lengths in comparison to longer lengths [2].

5.3.3   Error Rates and Types

In comparison with Sanger sequencing, the next-generation sequencing technologies have higher rates of errors of various types. The most common error type during next-generation sequencing is substitution (Chap. 4, Table 4.4), where a certain nucleotide is substituted with another type of nucleotide, making the identification of SNP even more difficult [11]. The Indel (insertion deletion) error is another common type of error in the Ion Torrent and SOLiD next-generation sequencing platforms (Chap. 4, Table 4.4). In certain cases, error types appear to be platform-specific as the source of the error is technically related to the technology implemented in a particular sequencing platform. For instance, CG deletion errors are exclusive to the PacBio RS platform (which will be described in more detail later) at a high rate (13 %) while A-T bias errors are specific to SOLiD sequencers but at a more average rate (Chap. 4, Table 4.4) [2, 12, 13]. Therefore, different next-generation platforms may possess varying error rates. This situation can complicate direct comparisons between the platforms due to the differing error types and the various templates utilized for assessments [2, 12, 13]. As a general rule, the error rate is known to increase when the maximum read length of the platform is approached. Hence, this factor complicates efforts to improve upon the read length [12].

Additional challenges in the next-generation field include complexities in library preparation and other procedural steps [14]. However, these will not be discussed in detail as the rapid pace of development in the field is known to overcome such technical challenges in due course. Thus, it is our expectation that most of today's technical challenges may be greatly diminished in the near future, especially with the introduction of third-generation sequencers. On the other hand, the associated computational challenges will likely become more complex in view of enhanced data generation from advancing sequencing methodology [2]. Therefore, we will focus on the discussion of such computational challenges and proposals to overcome them in the following chapters.

References

1. Mardis ER (2013) Next-generation sequencing platforms. Annu Rev Anal Chem (Palo Alto Calif) 6:287-303. doi:10.1146/annurev-anchem-062012-092628
2. El-Metwally S, Hamza T, Zakaria M, Helmy M (2013) Next-generation sequence assembly: four stages of data processing and computational challenges. PLoS Comput Biol 9 (12):e1003345. doi:10.1371/journal.pcbi.1003345
3. Schuster SC (2008) Next-generation sequencing transforms today's biology. Nat Methods 5 (1):16-18. doi:10.1038/nmeth1156
4. Gibson DG, Glass JI, Lartigue C, Noskov VN, Chuang RY et al. (2010) Creation of a bacterial cell controlled by a chemically synthesized genome. Science 329 (5987):52-56. doi:10.1126/science.1190719
5. Collins FS, Hamburg MA (2013) First FDA authorization for next-generation sequencer. N Engl J Med 369 (25):2369-2371. doi:10.1056/NEJMp1314561
6. Illumina Inc. (2014) HiSeq X Ten. http://www.illumina.com/systems/hiseq-x-sequencing-system.ilmn. Accessed 17-01-2014
7. WHO.int (2005) Genetics, genomics and the patenting of DNA: Review of potential implications for health in developing countries. 2005. http://www.who.int/genomics/FullReport.pdf. Accessed 10-01-2014
8. WHO.int (2013) Wild poliovirus in the Horn of Africa – update. http://www.who.int/csr/don/2013_10_01/en/. Accessed 10-01-2014
9. Liu L, Li Y, Li S, Hu N, He Y et al. (2012) Comparison of next-generation sequencing systems. J Biomed Biotechnol 2012:251364. doi:10.1155/2012/251364
10. Quail MA, Smith M, Coupland P, Otto TD, Harris SR et al. (2012) A tale of three next generation sequencing platforms: comparison of Ion Torrent, Pacific Biosciences and Illumina MiSeq sequencers. BMC Genomics 13:341. doi:10.1186/1471-2164-13-341
11. Bansal V, Harismendy O, Tewhey R, Murray SS, Schork NJ et al. (2010) Accurate detection and genotyping of SNPs utilizing population sequencing data. Genome Res 20 (4):537-545. doi:10.1101/gr.100040.109
12. Glenn TC (2011) Field guide to next-generation DNA sequencers. Mol Ecol Resour 11 (5):759-769. doi:10.1111/j.1755-0998.2011.03024.x
13. Glenn TC (2013) Field guide to next-generation DNA sequencers-Update. http://www.molecularecologist.com/next-gen-fieldguide-2013/. Accessed 10-01-2014
14. Oyola SO, Otto TD, Gu Y, Maslen G, Manske M et al. (2012) Optimizing Illumina next-generation sequencing library preparation for extremely AT-biased genomes. BMC Genomics 13:1. doi:10.1186/1471-2164-13-1

Chapter 6
New Horizons in Next-Generation Sequencing

Abstract In the previous chapters, we described the most common and well-established next-generation sequencing technologies and platforms. However, several methodologies and sequencers with outstanding features have also been released in the last few years. Furthermore, additional technologies demonstrating great promise are currently in development. In this chapter, we will briefly describe these recent and ongoing developments that may have a profound impact on the future of sequencing.

6.1 Third-Generation Sequencing Methods

Despite the advantages of next-generation sequencing methods, soaring expectations in the field have driven the demand for even better technologies (see Chap. 5). Therefore, a new staple of sequencing methods known as third-generation sequencing or next-generation sequencing is being developed in the hopes of elevating the platform to a whole new dimension [1]. Ideally, third-generation sequencing methodology should reduce or eliminate some or all of the three main challenges faced by the next-generation techniques, i.e., excessive machine costs, short read lengths, and significant error rate. To date, three methods have been introduced that can be considered as third-generation methods or in the transitionary phase between the next-generation and third-generation tools.

6.1.1 Heliscope Single-Molecule Sequencing

Heliscope Single-Molecule Sequencing (or Helicos Single-Molecule Fluorescent Sequencing) is the first single-molecule sequencing (SMS) method that can directly identify the exact sequence of a given DNA stretch [2]. In this technique, the DNA

S. El-Metwally et al., *Next Generation Sequencing Technologies and Challenges in Sequence Assembly*, SpringerBriefs in Systems Biology 7, DOI 10.1007/978-1-4939-0715-1_6, © The Authors 2014

to be sequenced is sheared and the resulting fragments are then attached to Poly-A tails, which allow the fragments to be connected to a flow cell surface. A single type of fluorescently labeled nucleotide is added in cycles to extend the DNA by one nucleotide per cycle. After the addition of each nucleotide, the reaction is paused using a terminating nucleotide in order to capture an image of the florescent label. Subsequently, the flow cell surface is washed and the blocking is removed to repeat the cycle [3]. This technology was developed by Helicos Biosciences and was used in 2009 to sequence whole human genome (the genome of Stephen Quake, Professor of Stanford University, USA and a co-founder of Helicos BioSciences) for less than 50,000 dollars [2]. It was also used to sequence the genome of the M13 bacteriophage [4]. However, by the end of 2012, Helicos BioSciences shut its doors and filed for bankruptcy.

6.1.2   Single-Molecule Real-Time Sequencing

The single-molecule real-time (SMRT) sequencing technique is another SMS method that is based on the principle of sequencing by synthesis. It utilizes small well-like containers with a single DNA polymerase enzyme affixed at the bottom of a structure called the zero-mode waveguide (ZMW) [5]. Each ZMW contains a polymerase enzyme and a DNA fragment as a template, and creates an observation volume that is sufficiently illuminated to view a single nucleotide when being incorporated by DNA polymerase. This observation is accomplished through capturing the florescent label of the incorporated nucleotide by a detector [6]. The SMRT Sequencing technology was developed by Pacific Biosciences and is currently implemented in their commercial sequencing machines, where the actual sequencing is fulfilled on a chip that contains several ZMVs (see below).

6.1.3   Nanopore Sequencing

The Nanopore sequencing method was first introduced in the middle of the 1990s as a technique for determining the nucleotide order in a DNA sequence [7]. The technique is based on the utilization of a surface comprising of 1 nm diameter pores. The passage of DNA through a pore alters its ion current. This effect is indicative of the types of nucleotides present as current changes depend on the shape, size, and length of the DNA molecules being sequenced. Thus, each nucleotide can be identified based on its corresponding ion blockage time. Nanopore sequencing is a promising and low-cost method that does not require modified nucleotides, chemical labeling, or PCR amplification [8].

The major challenge of utilizing the nanopore method is the preparation involved in developing the nanopore surface, which can be either solid-state nanopore surfaces or protein-based nanopore surfaces. Solid-state surfaces are used in solid-state nanopore sequencing techniques such as sequencing with florescent labels [9]. On the other hand, protein-based nanopore sequencing employs proteins such as *Alpha hemolysin* and *Mycobacterium smegmatis porin A* (MspA) as nanopore surfaces [10–12]. Nanopore sequencing is still in the developmental stages, and thus far have not been commercially available [13, 14].

6.2 Third-Generation Sequencing Platforms

6.2.1 HeliScope Single-Molecule Sequencer

The Heliscope Single-Molecule Sequencer was the first commercialized SMS developed by Helicos Biosciences in 2009. It implements the Heliscope SMS technology that was developed by the same company and represents a revolutionary sequencing paradigm that allows the sequencing of about one billion molecules in about 7 days, a rate 1,000-fold over the technology available when first released [2]. It uses novel reagents that allow digital measurement of homopolymer sequences as well as a new alignment algorithm to perform whole genome assembly (reference-based assembly). The sequencer reads are between 24 and 70 bp, which are very short based on previous expectations from a third-generation product. However, the higher speed of sequencing and lower associated costs are the significant strengths of the platform.

The Heliscope Single-Molecule Sequencer was used to sequence the genome of one of the co-founders of Helicos Biosciences (referred to as Patient Zero or P0 in the published article), with promising results [2]. Four sequencers were used to sequence the whole human genome and the results were mapped to ~90 % of the reference genome with a coverage depth near a Poisson distribution [2]. However, Helicos Biosciences closed down at the end of 2012 and, therefore, the Heliscope Single-Molecule Sequencer was excluded from comparisons in this chapter.

6.2.2 PacBio RS II

PacBio RS is a DNA sequencing system developed by Pacific Biosciences. The PacBio RS systems (PacBio RS and PacBio RS II) are single-molecule sequencers that implement the SMRT sequencing technology developed by the same company. These can be considered as genuine third-generation sequencers with a read length that is >3,000 bp, which is one of the longest available read lengths to date. The sequencer is compact with a short run time (~10 h). However, it is very expensive and still suffers from high error rates and a low total number of reads per run (Tables 6.1, 6.2, 6.3, and 6.4) [13, 14].

Table 6.1 Comparison of major third-generation sequencers advantages and disadvantages[a]

Instrument	Primary advantages	Primary disadvantages
Oxford Nanopore GridION 2000	Extremely long reads are feasible; low-cost instrument (node); nodes can be placed in standard computer racks; error rate doesn't increase along the length of the read; hairpin on one end allows reading of the complementary strand	Not yet available; no data publicly available; 4 % error rates; errors are likely to be biased (thus multiple reads will lead to higher confidence in the wrong answer)
Oxford Nanopore GridION 8000	Same as GridION 2000, but more reads per unit time; lower cost per GB	Same as GridION 2000
Oxford Nanopore minION	No Instrument; IT IS A USB DEVICE; can load "raw" samples	Not yet available; no data publicly available; high cost per MB relative to other Nanopore sequencers
PacBio	Single-molecule real-time sequencing; longest available read length; ability to detect base modifications; short instrument run time; random error profile; modest cost per sample; many methods being developed	High error rates; low total number of reads per run; high cost per MB; high capital cost; many methods still in development; weak company performance

[a] The data is derived from 2013 NGS Field Guide update [13, 33]

Table 6.2 Comparison of major third-generation sequencers run time, read length, and output data[a]

Instrument	Run time (h)	Millions of reads/run	Read length	Yield MB/run	Data file sizes (GB)
Oxford Nanopore GridION 2000	[NA]	[4]	[10,000]	[40,000]	[Variable]
Oxford Nanopore GridION 8000	[5]	[10]	[10,000]	[100,000]	[Variable]
Oxford Nanopore minion	≤6	[0.1]	[9,000]	1,000	[Small]
PacBio RS	≤2	0.03	>3,000	100–150	2 (basecalls, QV, kinetics)

[a] The data is derived from 2013 NGS Field Guide update [13, 33]

Table 6.3 Comparison of major third-generation sequencers purchase and operation costs[a]

Instrument	Reagent cost/run	Reagent cost/MB	Minimum unit cost (% run)	Purchase cost	Service contract	Computational resources
Oxford Nanopore GridION 2000	Varies	[$0.04]	–	–	–	–
Oxford Nanopore GridION 8000	Varies	$0.02	–	–	–	–
Oxford Nanopore minion	≤$900	$1	~$1,100 (10 %)	–	–	Laptop
PacBio RS	≥$300	$2–17	$500 (100 %)	$695	85	$65 cluster

[a] The data is derived from 2013 NGS Field Guide update [13], some values are from 2011's guide [33]. All costs are in thousands of US dollars

Table 6.4 Comparison of major next-generation sequencers errors and error rates[a]

Instrument	Data file sizes (GB)	Primary errors	Single-pass error rate (%)	Final error rate (%)
Oxford Nanopore GridION 2000	[Variable]	Deletions	≥4[a]	4[a]
Oxford Nanopore GridION 8000	[Variable]	Deletions	≥4[a]	4[a]
Oxford Nanopore minion	[Small]	Deletions	≥4[a]	4[a]
PacBio RS	2 (basecalls, QV, kinetics)	Indel	~13	≤1

[a] The data is derived from 2013 NGS Field Guide update [13, 33]

6.2.3   Oxford Nanopore GridION

The Oxford Nanopore GridION sequencers are sequencing machines that implement the Nanopore sequencing methodology. The sequencers are being developed by Oxford Nanopore Technologies Ltd. (UK), which had originally announced that their first commercialized instrument would be available by the end of 2013 [15]. However, at the time of manuscript preparation, it had not yet launched.

The Oxford Nanopore GridION systems promise small, inexpensive and high-throughput sequencers with an unprecedented long read length of ~10,000 bp. According to the product page on the company website [15], the Oxford Nanopore GridION can be used as a single desktop machine or stacked in racks in a similar manner to computer servers. Furthermore, it is stated that the instrument does not require a dedicated server and utilizes a single-use disposable cartridge that contains all the reagents necessary for the experiment. The available information on the performance of the Oxford Nanopore GridION systems shows a relatively high error rate (~4 %), though this rate does not rise upon increasing the read length [13, 14].

6.3   Sequencing Methods Under Development

We have previously discussed the rapid rate at which methodology has been developed in the DNA sequencing field, and how this fact has helped alleviate prior technical challenges. Moreover, several additional methods are currently in development and hold the promise of making DNA sequencing cheaper, easier, faster, and more accurate. The ultimate goal of these developments is to make whole human genome DNA sequencing as simple and affordable as other standard laboratory procedures. This would allow its widespread utilization towards innumerable clinical applications such as personalized medicine, and would augment research to unprecedented levels [16]. In this section, we will discuss methodologies that are presently in the developmental phase as well as their expected outcomes.

6.3.1 Solution-Based Hybridization Sequencing

The idea behind sequencing by hybridization is not a new one and has been previously presented [17]. Sequencing by hybridization involves a nonenzymatic approach based on the creation of a hybrid between the DNA molecule of interest and another molecule of known sequence. When one short strand of DNA binds to its complementary strand, the binding become very sensitive to mismatches, even at the level of a single-base. Thus, the sequence of the complementary strand can be inferred from the sequence of its hybrid. The method requires a library of DNA probes (short single-stranded DNA sequences) based on the organism of interest, its variants or its single-base variations, and can be accomplished using DNA chips or microarrays [17]. The technique has several advantages including homogenous coverage, though the preliminary requirement of DNA and the need for a significant amount of chemicals limit its overall utility. However, the recent introduction of solution-based hybridization has drastically reduced the dependency on chemicals and expensive equipment [18, 19].

6.3.2 Tunneling Current DNA Sequencing

The novel approach of identifying a DNA sequence and differentiating between the four types of nucleotides through the use of electrical signals was first presented via nanopore sequencing [8]. Based on these findings, the Tunneling Current DNA Sequencing method identifies specific nucleotides through tunneling current conducted by single-base molecules as they pass through a channel comprising of a pair of nanoelectrodes [10, 20, 21]. The differing structures of the nucleotides have varied effects on the current during this process. Thus, differentiating between them is possible through the identification of the characteristic changes in the current influenced by each nucleotide. A recent report also presented a hybrid method that combined single-base electrical identification and random sequencing to allow successful sequence reads from nine different DNA oligomers and microRNA [21]. The method promises an elevated sequencing speed in comparison to those currently available.

6.3.3 Microscopy-Based DNA Sequencing

Microscopy-based DNA Sequencing utilizes an electron microscope to directly visualize the nucleotide sequence of intact DNA molecules. In this approach, nucleotides are enzymatically modified to contain atoms with higher atomic number that can be directly visualized and identified by the electron microscope. Using this technique, an intact synthetic molecule of length >3,200 bp and an intact viral DNA of length >7,000 bp were sequenced successfully, proving the potential of this methodology in the sequencing of long intact DNA molecules [22].

6.3.4 Mass Spectrometry-Based DNA Sequencing

Mass Spectrometry is well known as the technology of choice in the study of proteins and the identification of amino acid sequences [23]. Additionally, it is utilized in the study of metabolites via the capillary electrophoresis mass spectrometry (CE-MS) approach [24]. For the purposes of DNA sequencing, electrospray ionization time-of-flight mass spectrometry (ESI-TOF MS) and matrix-assisted laser desorption ionization time-of-flight mass spectrometry (MALDI-TOF MS) were used to determine the nucleotide sequence of DNA through the examination of nucleotide mass. This contrasted with previous methodology that employed the study of nucleotide size, structure, florescent labeling, or radioactive labeling [25, 26]. Since each type of nucleotide has its own unique chemical structure, each of them also possesses a unique mass. Therefore, spectrometry can be used to identify the nucleotide sequences accurately and in high resolution. This method was found to be more effective with RNA, so the DNA is converted to RNA prior to the sequencing process. An early attempt to use MS for DNA sequencing showed that the longest read in the procedure could be 100 bp [27]. In more recent studies, MS-based DNA sequencing has been used to identify SNPs in pathogens [26] and the comparison of human mitochondrial DNA with DNA from the bones of dead soldiers during a forensic investigation [28].

6.3.5 RNA Polymerase Sequencing

RNA polymerase (RNAP) Sequencing involves the utilization of an RNAP enzyme that is attached to a polystyrene bead while the DNA molecule to be sequenced is attached to another bead, following which the two beads are placed in optical traps. The sequencing information is obtained from the movement of the nucleic acid enzyme and the sensitivity of the optical trap. During transcription, the motion of the RNAP brings the two beads closer, which can be recorded in single nucleotide resolution (in Angstrom range). The differentiation between the four types of nucleotides is then accomplished using a Sanger approach-like method. The concentration displacement of the four types of nucleotides over the transcription time is compared and used to pinpoint the specific types of the nucleotides in the sequence [29, 30].

In addition to the above, several other sequencing methods and instruments are currently either in the research phase or at the initial stages of commercialization. These include in vitro virus high-throughput sequencing [31] and microfluidic Sanger sequencing [32], for instance. However, due to text limitations, it is not possible to discuss them all within the confines of this book. Reports that survey or compare upcoming methods and platforms are readily available [13, 30], though the rapid pace of the field necessitates sources that are frequently updated such as the NGS Field Guide [33].

References

1. Rusk N (2009) Cheap third-generation sequencing. Nature Methods 6 (4):244-245. doi:10.1038/nmeth0409-244a
2. Pushkarev D, Neff NF, Quake SR (2009) Single-molecule sequencing of an individual human genome. Nature Biotechnology 27 (9):847-850. doi:10.1038/Nbt.1561
3. Thompson JF, Steinmann KE (2010) Single molecule sequencing with a HeliScope genetic analysis system. Curr Protoc Mol Biol Chapter 7:Unit7 10. doi:10.1002/0471142727. mb0710s92
4. Harris TD, Buzby PR, Babcock H, Beer E, Bowers J et al. (2008) Single-molecule DNA sequencing of a viral genome. Science 320 (5872):106-109. doi:10.1126/science.1150427
5. Levene MJ, Korlach J, Turner SW, Foquet M, Craighead HG et al. (2003) Zero-mode wave-guides for single-molecule analysis at high concentrations. Science 299 (5607):682-686. doi:10.1126/science.1079700
6. Eid J, Fehr A, Gray J, Luong K, Lyle J et al. (2009) Real-time DNA sequencing from single polymerase molecules. Science 323 (5910):133-138. doi:10.1126/science.1162986
7. Kasianowicz JJ, Brandin E, Branton D, Deamer DW (1996) Characterization of individual polynucleotide molecules using a membrane channel. Proc Natl Acad Sci U S A 93 (24):13770-13773
8. Schadt EE, Turner S, Kasarskis A (2010) A window into third-generation sequencing. Hum Mol Genet 19 (R2):R227-240. doi:10.1093/hmg/ddq416
9. McNally B, Singer A, Yu Z, Sun Y, Weng Z et al. (2010) Optical recognition of converted DNA nucleotides for single-molecule DNA sequencing using nanopore arrays. Nano Lett 10 (6):2237-2244. doi:10.1021/nl1012147
10. Stoddart D, Heron AJ, Mikhailova E, Maglia G, Bayley H (2009) Single-nucleotide discrimination in immobilized DNA oligonucleotides with a biological nanopore. Proc Natl Acad Sci U S A 106 (19):7702-7707. doi:10.1073/pnas.0901054106
11. Purnell RF, Mehta KK, Schmidt JJ (2008) Nucleotide identification and orientation discrimination of DNA homopolymers immobilized in a protein nanopore. Nano Lett 8 (9):3029-3034. doi:10.1021/nl802312f
12. Stoddart D, Maglia G, Mikhailova E, Heron AJ, Bayley H (2010) Multiple base-recognition sites in a biological nanopore: two heads are better than one. Angew Chem Int Ed Engl 49 (3):556-559. doi:10.1002/anie.200905483
13. Glenn TC (2011) Field guide to next-generation DNA sequencers. Mol Ecol Resour 11 (5):759-769. doi:10.1111/j.1755-0998.2011.03024.x
14. Glenn TC (2013) Field guide to next-generation DNA sequencers-Update. http://www. molecularecologist.com/next-gen-fieldguide-2013/. Accessed 10-01-2014
15. Oxford Nanopore Technologies Ltd. (2014) The GridION System. https://www.nanoporetech. com/technology/the-gridion-system/the-gridion-system. Accessed 10-01-2014
16. Collins FS, Hamburg MA (2013) First FDA authorization for next-generation sequencer. N Engl J Med 369 (25):2369-2371. doi:10.1056/NEJMp1314561
17. Hanna GJ, Johnson VA, Kuritzkes DR, Richman DD, Martinez-Picado J et al. (2000) Comparison of sequencing by hybridization and cycle sequencing for genotyping of human immunodeficiency virus type 1 reverse transcriptase. J Clin Microbiol 38 (7):2715-2721
18. Morey M, Fernandez-Marmiesse A, Castineiras D, Fraga JM, Couce ML et al. (2013) A glimpse into past, present, and future DNA sequencing. Mol Genet Metab 110 (1-2):3-24. doi:10.1016/j.ymgme.2013.04.024
19. Qin Y, Schneider TM, Brenner MP (2012) Sequencing by hybridization of long targets. PLoS One 7 (5):e35819. doi:10.1371/journal.pone.0035819
20. Di Ventra M (2013) Fast DNA sequencing by electrical means inches closer. Nanotechnology 24 (34):342501. doi:10.1088/0957-4484/24/34/342501
21. Ohshiro T, Matsubara K, Tsutsui M, Furuhashi M, Taniguchi M et al. (2012) Single-molecule electrical random resequencing of DNA and RNA. Sci Rep 2:501. doi:10.1038/srep00501

22. Bell DC, Thomas WK, Murtagh KM, Dionne CA, Graham AC et al. (2012) DNA base identification by electron microscopy. Microsc Microanal 18 (5):1049-1053. doi:10.1017/ S1431927612012615

23. Helmy M, Tomita M, Ishihama Y (2012) Peptide identification by searching large-scale tandem mass spectra against large databases: bioinformatics methods in proteogenomics. Genes Genome Genomics 6:76-85

24. Ishii N, Nakahigashi K, Baba T, Robert M, Soga T et al. (2007) Multiple high-throughput analyses monitor the response of *E. coli* to perturbations. Science 316 (5824):593-597. doi:10.1126/science.1132067

25. Edwards JR, Ruparel H, Ju J (2005) Mass-spectrometry DNA sequencing. Mutat Res 573 (1-2):3-12. doi:S0027-5107(05)00023-0

26. Beres SB, Carroll RK, Shea PR, Sitkiewicz I, Martinez-Gutierrez JC et al. (2010) Molecular complexity of successive bacterial epidemics deconvoluted by comparative pathogenomics. Proc Natl Acad Sci U S A 107 (9):4371-4376. doi:10.1073/pnas.0911295107

27. Monforte JA, Becker CH (1997) High-throughput DNA analysis by time-of-flight mass spectrometry. Nat Med 3 (3):360-362

28. Howard R, Encheva V, Thomson J, Bache K, Chan YT et al. (2013) Comparative analysis of human mitochondrial DNA from World War I bone samples by DNA sequencing and ESI-TOF mass spectrometry. Forensic Sci Int Genet 7 (1):1-9. doi:10.1016/j.fsigen.2011.05.009

29. Greenleaf WJ, Block SM (2006) Single-molecule, motion-based DNA sequencing using RNA polymerase. Science 313 (5788):801. doi:313/5788/801

30. Pareek CS, Smoczynski R, Tretyn A (2011) Sequencing technologies and genome sequencing. J Appl Genet 52 (4):413-435. doi:10.1007/s13353-011-0057-x

31. Fujimori S, Hirai N, Ohashi H, Masuoka K, Nishikimi A et al. (2012) Next-generation sequencing coupled with a cell-free display technology for high-throughput production of reliable interactome data. Sci Rep 2:691. doi:10.1038/srep00691

32. Chen YJ, Roller EE, Huang X (2010) DNA sequencing by denaturation: experimental proof of concept with an integrated fluidic device. Lab Chip 10 (9):1153-1159. doi:10.1039/b921417h

33. Gurevich A, Saveliev V, Vyahhi N, Tesler G (2013) QUAST: quality assessment tool for genome assemblies. Bioinformatics 29 (8):1072-1075. doi:10.1093/bioinformatics/btt086

Chapter 7
Novel Next-Generation Sequencing Applications

Abstract Next-generation sequencing technologies have pushed the envelope beyond the primary goal of identifying the sequence of nucleotides within a given DNA molecule to a whole new multitude of applications. In this chapter, we describe select novel applications of next-generation sequencing in relation to large-scale sequencing-based projects, cell and cell compartments sequencing and disease-targeted sequencing.

7.1 Introduction

The applications of the next-generation sequencing technologies and the recently introduced third-generation sequencing methodologies are nearly limitless. The determination of the constituents of a DNA sequence itself was the primary aim of the first-generation of sequencing methods. With the availability of next-generation technologies, sequencing of the genome went from being a research aim to an important discovery tool. Thus, the utilization of whole genome sequencing (WGS), which is the primary application of these technologies, experienced a remarkable growth in the last few years. For instance, the number of genome sequencing projects in the Genome Online Database (GOLD) increased from 10,420 projects in May 2011 to 37,540 projects in January 2014 [1, 2]. The completed and published genomes in the above periods were 1,700 and 12,720 genomes, respectively, which demonstrated an incredible 720 % increase in a span of just 3 years. Clearly, this increase reflects the improved availability, affordability, and efficiency of the existing sequencers and methods.

The increased ease at which genome sequencing could be acquired opened the floodgates for applications and discoveries that went well beyond the initial goals of identifying the order of nucleotides or gene structure. Next-generation sequencing is presently being used in the WGS of humans [3], animals [4, 5], plants [6, 7], microbes [8, 9] and viruses [10]. In addition to WGS, next- and third-generation

S. El-Metwally et al., *Next Generation Sequencing Technologies*
and Challenges in Sequence Assembly, SpringerBriefs in Systems Biology 7,
DOI 10.1007/978-1-4939-0715-1_7, © The Authors 2014

sequencing technologies are also employed in genome resequencing [11], RNA sequencing (RNA-seq) [12], whole exome sequencing (WES) [13], targeted sequencing [14], single-nucleotide variations discovery, analysis and validation [15], chromatin immunoprecipitation sequencing (ChIP-seq) [16], epigenetics [17], proteogenomics [18, 19], diseases and disorders targeted sequencing [20], mutations discovery [21], cancer research [22, 23] and numerous other clinical and health applications [24]. Several reports have extensively reviewed the genome sequencing applications [25, 26]. Furthermore, Nature Reviews Genetics has dedicated an ongoing article series to the applications of next-generation sequencing since 2009 [27]. Here, we will focus on the discussion of select novel applications that have been approached on a radically different scale since the advent of newer sequencing technologies.

7.2 Large-Scale Applications

Scientists design their research projects based on the availability and affordability of research tools and technologies. Thus, the availability of faster, cheaper, and more accurate tools and technologies leads to the planning of projects at an even higher level. The developments in genome sequencing technologies over the last decade have led to massive strides in sequencing power at an affordable cost and within a reasonable timeframe. This success has encouraged more expansive research projects where next-generation sequencing is used as a tool to discover diversities among individuals within large populations and to understand the fundamentals of life and biological systems. Here, we will take a few examples of large-scale genome projects that only became possible through the inception of next-generation sequencing and its subsequent development.

7.2.1 Genome 10K Project

In the year 2009, a group of genomics scientists established the Genome 10K Community of Scientists (G10KCOS) and announced the Genome 10K Project [28, 29]. The Genome 10K Project aims to sequence and annotate the genomes of about 10,000 vertebrate species that will amount to almost one species from each vertebrate genus. The project was inspired by the human genome project and the subsequent availability of 56 vertebrate (32 mammals and 24 nonmammalian) genomes that are appropriate for comparative genomic analyses [29]. The stated timeframe for the project is quite short as the community aims to assemble such a "genomic zoo" by 2015. The targeted species are distributed between all the vertebrates, including mammals, birds, non-avian reptiles, amphibians, and fishes. After 1 year, the G10KCOS announced the first 101 species to be sequenced [30]. Since fishes represent more than 50 % of extant vertebrates, the Genome 10K Project intends to sequence the genomes of about 4,000 fish species, 160 of which are currently in progress [31].

It is likely that the Genome 10K Project may take longer than expected. Nevertheless, it is a staggering effort that promises unique comparative study opportunities that is only possible through the use of modern genome sequencing technologies.

7.2.2 Tree of Life Sequencing Project

Another example of large-scale genome sequencing projects is the Tree of Life Sequencing Project that was announced by Beijing Genome Institute (BGI) in 2010. BGI has the most powerful sequencing capacity worldwide, and is the main contributor to the 1000 Genomes and Genome 10K projects as well. The Tree of Life Sequencing Project is also known as the 1000 Plant & Animal Reference Genomes Project, a name that is more descriptive of the intended goals of the venture. The project aims to target 1,000 reference genomes from 500 animals and 500 plants of various economically and scientifically important species such as rice, silkworm, cucumber, panda, camel, oyster, ant, grouper, goose, crested ibis, and potato genomes. To date, 106 genomes have been completed and published while another 200 are in progress, representing about 30 % of the targeted species [32].

7.3 Cell and Cell Compartments Applications

The projects discussed in the previous section shared the tendency to sequence a huge number of organisms and provide their genomes as reference genomes. In contrast, we will now examine the application of next-generation sequencing on a much smaller scale, such as a single cell or even a cell compartment. The main aims of such applications are to sequence the genomes of species that are difficult to grow in the lab environment, or when the availability of samples is limited. Another interesting possibility is the determination of the heterogeneity between single cells in normal or tumorous tissues.

7.3.1 Single-Cell Genome Sequencing

Preparation of sequencing samples is initiated with a group of cells, e.g., cell cultures of bacteria or archaea. However, culturing attempts have failed in the case of several microorganisms, making full genome sequencing of such organisms unlikely [33, 34]. Thus, methods to sequence a single cell were developed using PCR-based amplification of the single bacterial cell genome with accuracy approaching 97 % [35]. Another technique that increased accuracy to 99.6 % [36] involved PCR-based amplification with multiple displacement amplification (MDA) [34] followed by post-amplification normalization and assembly with the reference genome. These methods can be used to sequence the genomes of either single cells or individual cells from a variety of samples (with different treatments or from different environmental sources).

Single-cell sequencing applications have also expanded to include the study of diseases with genetic alterations and to find variations (heterogeneity) between different cells in diseased tissue. For instance, the amount of cancer-related genomic mutations in the Catalogue Of Somatic Mutations In Cancer (COSMIC) database number over one million to date [37]. The heterogeneity of tumor cells can result in several complications such as developing rare chemo-resistant cells that can resist chemotherapy. Such cells can regrow and result in the formation of a chemo-resistant tumor [38]. Several attempts have been made to apply single-cell sequencing to cancer genomics, allowing the possibility to sequence up to 200 single cells independently during a single run [39]. The numerous single-cell sequencing applications in cancer can include the pinpointing of chemo-resistant cells, the early detection of tumor cells, measuring intratumor heterogeneity, monitoring of circulating tumor cells (CTCs) and in drug target discovery [40, 41]. Furthermore, the techniques may also be utilized to develop a guided form of chemotherapy that is appropriate against the measured heterogeneity of the tumor [39]. In the later sections, we will discuss further details on the applications of sequencing in cancer.

7.3.2 Mitochondrial Genome Sequencing

Mitochondria are cellular organelles that can be found in eukaryotic cells. They are responsible for producing most of the cell's energy by supplying it with adenosine triphosphate (ATP) through the phosphorylation of adenosine diphosphate (ADP). Mitochondria have their own genome and genetics that are independent from the cell nucleus genome. Therefore, it has its own proteome that is about 615 proteins [41]. Most of the mitochondria are inherited from the mother, and there is group of diseases known as mitochondrial diseases that are caused by dysfunctional mitochondria or genes that are inherited through the mitochondrial genome [42, 43]. These structures are also attributed to play an important role in aging and cancer [44, 45]. Moreover, they have a special genetic code for tryptophan and methionine as well as a distinct stop codon. This allows the mitochondrial genome to be perfectly suited for forensic investigations and human phylogenic studies [44, 45]. Hence, advancements in next-generation sequencing [43, 46] have been aptly reflected in the utility of human mitochondrial genome sequencing during forensic investigations and cancer [45, 46] as well as the study of plants [47] and fish [48].

7.4 Disease-Targeted Sequencing

Several diseases are associated with genetic mutations or genetic disorders while others are inherited from carrier parents to their offspring. The ongoing discovery of disease-related genes has made disease-targeted sequencing tests an important diagnostic tool [49]. With Sanger sequencing, tests were designed for diseases with a

Table 7.1 Clinically available disease-targeted tests[a]

Disease area	Disease type	Number of genes
Cancer	Hereditary cancers (for example, breast, colon, and ovarian)	10–50
Cardiac diseases	Cardiomyopathies	50–70
	Arrhythmias (for example, long QT syndrome)	10–30
	Aortopathies (for example, Marfan's syndrome)	10
Immune disorders	Severe combined immunodeficiency syndrome	18
	Periodic fever	7
Neurological, neuromuscular and metabolic disorders	Ataxia	40
	Cellular energetics, metabolism	656
	Congenital disorders of glycosylation	23–28
	Dementia (for example, Parkinson's disease and Alzheimer's disease)	32
	Developmental delay, autism, intellectual disability	30–150
	Epilepsy	53–130
	Hereditary neuropathy	34
	Microcephaly	11
	Mitochondrial disorders	37–450
	Muscular dystrophy	12–45
Sensory disorders	Eye disease (for example, retinitis pigmentosa)	66–140
	Hearing loss and related syndromes	23–72
Other	Rasopathies (for example, Noonan's syndrome)	10
	Pulmonary disorders (for example, cystic fibrosis)	12–40
	Short stature	12

[a] Data is derived from [49]

single causative gene in order to confirm the diagnosis. On the other hand, designing tests for diseases with enormous genetic heterogeneity is far more difficult [49]. With the introduction of next-generation sequencing, the significant increase in throughput and reduction in technical costs greatly aid the design of tests for a wide spectrum of diseases and genetic disorders, as well as the discovery of new disease-related genes and mutations (Table 7.1). In this section, we will introduce some of the recent applications of next-generation sequencing in understanding inherited and complex diseases, including the study of disease-related genes and mutations.

7.4.1   Sequencing in Cancer

Cancer is widely known to be associated with somatic mutations [22]. The Sanger Institute launched the Cancer Genome Project (CGP) as one of the earliest attempts to identify cancer genes and mutations [50]. The CGP currently represents one of the main resources of cancer genomics and mutations with its several databases and resources, including the COSMIC database [51], the Cancer Gene Census [52], COSMIC whole genomes and the COSMIC cell-line project [37]. To date, over a million identified mutations in cancer have been cataloged in the COSMIC database, including all types of known genetic mutations such as single-nucleotide mutations, insertions, deletions, and chromosomal rearrangements [37, 51]. Although the primary technology utilized at the commencement of the CGP was Sanger sequencing, the project also utilized the power of next-generation sequencing in later phases.

Several other large-scale projects have been conceived through international consortiums aided by public and private funding. These projects also aim to identify cancer-related mutations and genes as well as categorize findings based on importance and recurrence. For example, the International Cancer Genome Consortium (ICGC) is a huge publicly funded cancer genome-sequencing project. The ICGC aims to sequence the whole genome of 50 different types and subtypes of cancer that are clinically important [53]. The most recent data release from the project (Release 14) provides the results of 41 different cancer projects from over 8,500 donors. In this case, sequencing studies resulted in the identification of over two million mutations from 54,682 mutated genes.

With relation to privately funded projects, the Pediatric Cancer Genome Project (PCGP) was announced in 2010 by St. Jude Children's Research Hospital and the Genome Institute at Washington University [54]. This project targeted the sequencing of 600 pediatric tumors and matched non-tumor germline samples (totaling 1,600 genomes) with high resolution sequencing in an aim to catalog somatic mutations of pediatric tumors and define the major subtypes in pediatric cancers [54]. The most recent data release from the PCGP (June 2013) contained the whole genomes of 15 different cancer types from over 360 patients that were analyzed and revealed novel findings [55].

7.4.2   Sequencing in Inherited Human Diseases

Inherited human diseases are disorders that result from single-gene mutations. They are also known as monogenic disorders or Mendelian disorders. There are around 5,000 known monogenic disorders though the genetic causes of most of them are still unknown [56]. Most of these cases resulted from exonic mutations (mutations that occur in the exon) or splice-site mutations (mutations that affect the splicing pattern of the mRNA). Both types of mutations affect the resulting protein sequence following translation of the affected gene [57]. Thus, whole exome sequencing (WES) using next-generation sequencing is an efficient methodology to identify both these types of mutations without the need of whole genome sequencing (WGS). Furthermore, the utilization of WES saves time and reduces cost since the human exome represents ~1 % of the human genome. However, certain other mutations that cannot be identified without sequencing the whole genome may also result from deletions [57]. The 1,000 Mendelian Disorders Project is a large-scale effort at the Beijing Genome Institute (BGI) that aims to sequence the genome of 1,000 Mendelian disorders to identify the causative genes behind them using next-generation sequencing rather than traditional techniques such as positional cloning, physical mapping, and candidate-gene sequencing [56].

7.4.3 Sequencing in Complex Human Diseases

Next-generation sequencing has provided novel approaches in locating common and rare variants that influence the risk of developing complex diseases such as cancer, diabetes, cardiovascular disease, and psychiatric disorders [25]. Several Genome-Wide Association Studies (GWAS) have used next-generation sequencing technologies in examining complex trait genetics [58, 59]. Such studies demonstrated the utility of next-generation sequencing applications in understanding complex diseases such as hypertrophic cardiomyopathy [59], brain disease [60] and diabetes [61]. Moreover, the investigations provided novel insight into understanding the genetics mechanisms behind disorders of sex development (DSD) [62].

References

1. Kyrpides NC (1999) Genomes OnLine Database (GOLD 1.0): a monitor of complete and ongoing genome projects world-wide. Bioinformatics 15 (9):773-774. doi:btc112
2. Pagani I, Liolios K, Jansson J, Chen IM, Smirnova T et al. (2012) The Genomes OnLine Database (GOLD) v.4: status of genomic and metagenomic projects and their associated metadata. Nucleic Acids Res 40 (Database issue):D571-579. doi:10.1093/nar/gkr1100
3. Yamey G (2000) Scientists unveil first draft of human genome. BMJ 321 (7252):7
4. Michelizzi VN, Dodson MV, Pan Z, Amaral ME, Michal JJ et al. (2010) Water buffalo genome science comes of age. Int J Biol Sci 6 (4):333-349
5. Edwards CJ, Magee DA, Park SD, McGettigan PA, Lohan AJ et al. (2010) A complete mitochondrial genome sequence from a mesolithic wild aurochs (Bos primigenius). PLoS One 5 (2):e9255. doi:10.1371/journal.pone.0009255
6. Cao J, Schneeberger K, Ossowski S, Gunther T, Bender S et al. (2011) Whole-genome sequencing of multiple Arabidopsis thaliana populations. Nat Genet 43 (10):956-963. doi:10.1038/ng.911
7. Matsumoto T, Wu J, Antonio BA, Sasaki T (2008) Development in rice genome research based on accurate genome sequence. Int J Plant Genomics 2008:348621. doi:10.1155/2008/348621
8. Shendure J, Porreca GJ, Reppas NB, Lin X, McCutcheon JP et al. (2005) Accurate multiplex polony sequencing of an evolved bacterial genome. Science 309 (5741):1728-1732. doi:1117389
9. Otero JM, Vongsangnak W, Asadollahi MA, Olivares-Hernandes R, Maury J et al. (2010) Whole genome sequencing of Saccharomyces cerevisiae: from genotype to phenotype for improved metabolic engineering applications. BMC Genomics 11:723. doi:10.1186/1471-2164-11-723
10. Ren X, Yang F, Hu Y, Zhang T, Liu L et al. (2013) Full genome of influenza A (H7N9) virus derived by direct sequencing without culture. Emerg Infect Dis 19 (11):1881-1884. doi:10.3201/eid1911.130664
11. Costa GG, Cardoso KC, Del Bem LE, Lima AC, Cunha MA et al. (2010) Transcriptome analysis of the oil-rich seed of the bioenergy crop Jatropha curcas L. BMC Genomics 11:462. doi:10.1186/1471-2164-11-462
12. Wang H, Zou Z, Wang S, Gong M (2013) Global Analysis of Transcriptome Responses and Gene Expression Profiles to Cold Stress of Jatropha curcas L. PLoS One 8 (12):e82817. doi:10.1371/journal.pone.0082817
13. Lai CC, Yeh YH, Hsieh WP, Kuo CT, Wang WC et al. (2013) Whole-Exome Sequencing to Identify a Novel LMNA Gene Mutation Associated with Inherited Cardiac Conduction Disease. PLoS One 8 (12):e83322. doi:10.1371/journal.pone.0083322
14. Bazak L, Haviv A, Barak M, Jacob-Hirsch J, Deng P et al. (2013) A-to-I RNA editing occurs at over a hundred million genomic sites, located in a majority of human genes. Genome Res. doi:gr.164749.113

15. Rellstab C, Zoller S, Tedder A, Gugerli F, Fischer MC (2013) Validation of SNP allele frequencies determined by pooled next-generation sequencing in natural populations of a non-model plant species. PLoS One 8 (11):e80422. doi:10.1371/journal.pone.0080422
16. Marinov GK, Kundaje A, Park PJ, Wold BJ (2013) Large-Scale Quality Analysis of Published ChIP-seq Data. G3 (Bethesda). doi:g3.113.008680v1
17. Milavetz B, Kallestad L, Woods E, Christensen K, Gefroh A et al. (2013) Erratum: Transcription and replication result in distinct epigenetic marks following repression of early gene expression. Front Genet 4:259. doi:10.3389/fgene.2013.00259
18. Helmy M, Sugiyama N, Tomita M, Ishihama Y (2012) Mass spectrum sequential subtraction speeds up searching large peptide MS/MS spectra datasets against large nucleotide databases for proteogenomics. Genes Cells 17 (8):633-644. doi:10.1111/j.1365-2443.2012.01615.x
19. Helmy M, Tomita M, Ishihama Y (2012) Peptide identification by searching large-scale tandem mass spectra against large databases: bioinformatics methods in proteogenomics. Genes Genome Genomics 6:76-85
20. Clement JA, Toulza E, Gautier M, Parrinello H, Roquis D et al. (2013) Private Selective Sweeps Identified from Next-Generation Pool-Sequencing Reveal Convergent Pathways under Selection in Two Inbred Schistosoma mansoni Strains. PLoS Negl Trop Dis 7 (12):e2591. doi:10.1371/journal.pntd.0002591
21. Damerla RR, Chatterjee B, Li Y, Francis RJ, Fatakia SN et al. (2013) Ion Torrent sequencing for conducting genome-wide scans for mutation mapping analysis. Mamm Genome. doi:10.1007/s00335-013-9494-7
22. Helmy M, Sugiyama N, Tomita M, Ishihama Y (2010) Onco-proteogenomics: a novel approach to identify cancer-specific mutations combining proteomics and transcriptome deep sequencing. Genome Biol 11. doi:10.1186/Gb-2010-11-S1-P17
23. Patel L, Parker B, Yang D, Zhang W (2013) Translational genomics in cancer research: converting profiles into personalized cancer medicine. Cancer Biol Med 10 (4):214-220. doi:10.7497/j.issn.2095-3941.2013.04.005
24. Chang F, Li MM (2013) Clinical application of amplicon-based next-generation sequencing in cancer. Cancer Genet. doi:S2210-7762(13)00142-7
25. Pareek CS, Smoczynski R, Tretyn A (2011) Sequencing technologies and genome sequencing. J Appl Genet 52 (4):413-435. doi:10.1007/s13353-011-0057-x
26. Soon WW, Hariharan M, Snyder MP (2013) High-throughput sequencing for biology and medicine. Mol Syst Biol 9:640. doi:10.1038/msb.2012.61
27. Nature Reviews Genetics Article Serie (2009) Applications of next-generation sequencing. http://www.nature.com/nrg/series/nextgeneration/index.html. Accessed 10-01-2014
28. Genome 10K Project (2009) Genome 10K Project Home Page. https://genome10k.soe.ucsc.edu/. Accessed 10-01-2014
29. Genome 10K Scientists (2009) Genome 10K: a proposal to obtain whole-genome sequence for 10,000 vertebrate species. J Hered 100 (6):659-674. doi:10.1093/jhered/esp086
30. Wagman B (2010) Genome 10K project announces first 101 species for genome sequencing. http://cbse.soe.ucsc.edu/news/article/1820?ID=1820. Accessed 10-01-2014
31. Bernardi G, Wiley EO, Mansour H, Miller MR, Orti G et al. (2012) The fishes of Genome 10K. Mar Genomics 7:3-6. doi:10.1016/j.margen.2012.02.002
32. Li Q, Li N, Hu X, Li J, Du Z et al. (2011) Genome-wide mapping of DNA methylation in chicken. PLoS One 6 (5):e19428. doi:10.1371/journal.pone.0019428
33. Lasken RS (2012) Genomic sequencing of uncultured microorganisms from single cells. Nat Rev Microbiol 10 (9):631-640. doi:10.1038/nrmicro2857
34. Dean FB, Hosono S, Fang L, Wu X, Faruqi AF et al. (2002) Comprehensive human genome amplification using multiple displacement amplification. Proc Natl Acad Sci U S A 99 (8):5261-5266. doi:10.1073/pnas.082089499
35. Woyke T, Tighe D, Mavromatis K, Clum A, Copeland A et al. (2010) One bacterial cell, one complete genome. PLoS One 5 (4):e10314. doi:10.1371/journal.pone.0010314
36. Rodrigue S, Malmstrom RR, Berlin AM, Birren BW, Henn MR et al. (2009) Whole genome amplification and de novo assembly of single bacterial cells. PLoS One 4 (9):e6864. doi:10.1371/journal.pone.0006864

37. Forbes SA, Bindal N, Bamford S, Cole C, Kok CY et al. (2011) COSMIC: mining complete cancer genomes in the Catalogue of Somatic Mutations in Cancer. Nucleic Acids Res 39 (Database issue):D945-950. doi:10.1093/nar/gkq929

38. Oakman C, Santarpia L, Di Leo A (2010) Breast cancer assessment tools and optimizing adjuvant therapy. Nat Rev Clin Oncol 7 (12):725-732. doi:10.1038/nrclinonc.2010.170

39. Navin N, Hicks J (2011) Future medical applications of single-cell sequencing in cancer. Genome Med 3 (5):31. doi:10.1186/gm247

40. Spaethling JM, Eberwine JH (2013) Single-cell transcriptomics for drug target discovery. Curr Opin Pharmacol 13 (5):786-790. doi:10.1016/j.coph.2013.04.011

41. Campbell NA (ed) (2003) Biology: Exploring Life. Pearson Prentice Hall, Upper Saddle River, NJ

42. Dimauro S, Davidzon G (2005) Mitochondrial DNA and disease. Ann Med 37 (3):222-232. doi:R53R2R64844U5876

43. Lyons EA, Scheible MK, Sturk-Andreaggi K, Irwin JA, Just RS (2013) A high-throughput Sanger strategy for human mitochondrial genome sequencing. BMC Genomics 14:881. doi:10.1186/1471-2164-14-881

44. Wallace DC (2010) Mitochondrial DNA mutations in disease and aging. Environ Mol Mutagen 51 (5):440-450. doi:10.1002/em.20586

45. Brandon M, Baldi P, Wallace DC (2006) Mitochondrial mutations in cancer. Oncogene 25 (34):4647-4662. doi:1209607

46. Parson W, Strobl C, Huber G, Zimmermann B, Gomes SM et al. (2013) Reprint of: Evaluation of next generation mtGenome sequencing using the Ion Torrent Personal Genome Machine (PGM). Forensic Sci Int Genet 7 (6):632-639. doi:10.1016/j.fsigen.2013.09.007

47. Fajardo D, Schlautman B, Steffan S, Polashock J, Vorsa N et al. (2013) The American cranberry mitochondrial genome reveals the presence of selenocysteine (tRNA-Sec and SECIS) insertion machinery in land plants. Gene. doi:S0378-1119(13)01648-X

48. Hester J, Atwater K, Bernard A, Francis M, Shivji MS (2013) The complete mitochondrial genome of the basking shark Cetorhinus maximus (Chondrichthyes, Cetorhinidae). Mitochondrial DNA. doi:10.3109/19401736.2013.845762

49. Rehm HL (2013) Disease-targeted sequencing: a cornerstone in the clinic. Nat Rev Genet 14 (4):295-300. doi:10.1038/nrg3463

50. Davies H, Bignell GR, Cox C, Stephens P, Edkins S et al. (2002) Mutations of the BRAF gene in human cancer. Nature 417 (6892):949-954. doi:10.1038/nature00766

51. Alexandrov LB, Nik-Zainal S, Wedge DC, Aparicio SA, Behjati S et al. (2013) Signatures of mutational processes in human cancer. Nature 500 (7463):415-421. doi:10.1038/nature12477

52. Futreal PA, Coin L, Marshall M, Down T, Hubbard T et al. (2004) A census of human cancer genes. Nat Rev Cancer 4 (3):177-183. doi:10.1038/nrc1299

53. Hudson TJ, Anderson W, Artez A, Barker AD, Bell C et al. (2010) International network of cancer genome projects. Nature 464 (7291):993-998. doi:10.1038/nature08987

54. Chen X, Stewart E, Shelat AA, Qu C, Bahrami A et al. (2013) Targeting oxidative stress in embryonal rhabdomyosarcoma. Cancer Cell 24 (6):710-724. doi:10.1016/j.ccr.2013.11.002

55. Zhang J, Wu G, Miller CP, Tatevossian RG, Dalton JD et al. (2013) Whole-genome sequencing identifies genetic alterations in pediatric low-grade gliomas. Nat Genet 45 (6):602-612. doi:10.1038/ng.2611

56. Liu Y, Gao M, Lv YM, Yang X, Ren YQ et al. (2011) Confirmation by exome sequencing of the pathogenic role of NCSTN mutations in acne inversa (hidradenitis suppurativa). J Invest Dermatol 131 (7):1570-1572. doi:10.1038/jid.2011.62

57. Kuhlenbaumer G, Hullmann J, Appenzeller S (2011) Novel genomic techniques open new avenues in the analysis of monogenic disorders. Hum Mutat 32 (2):144-151. doi:10.1002/humu.21400

58. Day-Williams AG, Zeggini E (2011) The effect of next-generation sequencing technology on complex trait research. Eur J Clin Invest 41 (5):561-567. doi:10.1111/j.1365-2362.2010.02437.x

59. Voelkerding KV, Dames S, Durtschi JD (2010) Next generation sequencing for clinical diagnostics-principles and application to targeted resequencing for hypertrophic cardiomyopathy: a paper from the 2009 William Beaumont Hospital Symposium on Molecular Pathology. J Mol Diagn 12 (5):539-551. doi:10.2353/jmoldx.2010.100043

60. Zoghbi HY, Warren ST (2010) Neurogenetics: advancing the "next-generation" of brain research. Neuron 68 (2):165-173. doi:10.1016/j.neuron.2010.10.015
61. Nejentsev S, Walker N, Riches D, Egholm M, Todd JA (2009) Rare variants of IFIH1, a gene implicated in antiviral responses, protect against type 1 diabetes. Science 324 (5925):387-389. doi:10.1126/science.1167728
62. Bashamboo A, Ledig S, Wieacker P, Achermann JC, McElreavey K (2010) New technologies for the identification of novel genetic markers of disorders of sex development (DSD). Sex Dev 4 (4-5):213-224. doi:10.1159/000314917

Part III
Next-generation Sequence Assembly Stages, Assessments, Tools and Challenges

Chapter 8
Next-Generation Sequence Assembly Overview

Abstract Next-generation sequence assembly can be viewed as a five-stage process of data processing and computational challenges. These stages are error correction, graph construction, graph simplification, scaffolding, and the assembly assessment stage. These stages communicate with each other to produce the final assembled sequences. Each stage receives a set of inputs from the preceding one and passes its output to the following stage. In this chapter, we will briefly introduce the basic functions of each stage and provide a coherent framework of the communications that occur between them.

8.1 Introduction to Next-Generation Sequence Assembly

The sequence assembly process was developed to resolve the limitations of current technologies that prevent the sequencing of the whole genome/chromosome during a single read. In first- and next-generation sequencing methods (see Chap. 3), the whole genome is sheared into short random fragments with short overlaps. Each fragment is sequenced independently and the resulting sequences are individually called a "read". Hence, the process of repositioning these random reads to reconstruct the whole genome is known as the "sequence assembly process" [1, 2].

According to the sample and type of raw data generated by sequencing instruments and the aim of the study, the assembly process may take many flavors including genome, transcriptome, or metagenome sequence assembly. If the raw data in the sequencing experiment is genomic DNA, the process is called genome assembly. Likewise, if the raw data is mRNA, the process is called transcriptome assembly, whereas assembling reads resulting from sequencing environmental samples that contain a mixture of organisms is called metagenome assembly. The ever-increasing number of applications in genomics, transcriptomics, metagenomics, and single-cell sequencing exhibits the need to acquire sequences from the viral, microbial, bacterial, or eukaryotic communities [3]. While the details of the assembly process

S. El-Metwally et al., *Next Generation Sequencing Technologies
and Challenges in Sequence Assembly*, SpringerBriefs in Systems Biology 7,
DOI 10.1007/978-1-4939-0715-1_8, © The Authors 2014

and employed assembly tools are different in each case, the sequence assembly process always shares the same stages.

The process of sequence assembly starts with filtering the reads to remove or correct errors and then computing a set of overlaps among them to discover their arrangement. These overlaps are used to connect the reads together into long contiguous structures called "contigs". Similarly, contigs can also be connected together to form even longer sequence stretches called "scaffolds" [4].

According to the availability of the reference sequences, the sequence assembly process has two main approaches, comparative sequence assembly and de novo sequence assembly. In comparative sequence assembly (also known as reference-based sequence assembly), reference sequences from the same organism or closely related species help to guide the reconstruction process [5]. On the other hand, de novo assembly does not involve reference sequences and consequently is a more complicated process [1].

8.2   Sequence Assembly Framework

Sequence assembly is a multiphase process. These phases communicate together in order to produce the final assembled sequence. Not only does the organization of these phases differ from one assembly to another, but some phases are completely missing in certain assembly processes in accordance with various issues (Fig. 8.1) [6].

The first phase, commonly known as the error correction phase, aims at filtering erroneous reads by removing or correcting sequencing errors. The filtered reads are

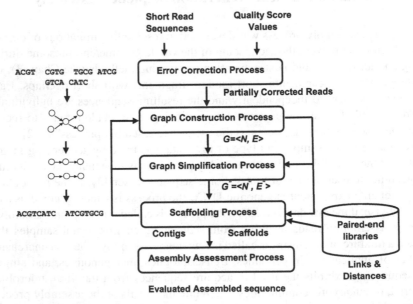

Fig. 8.1 Schematic representation of the five stages of next-generation sequence assembly process (*Note*: G'' is a repairing version of graph G with N nodes and E edges)

then fed into the second phase that formulates them into a graph of nodes with their relationships represented as graph edges. This representation overcomes the limitation of available computational resources that are necessary to manage the high throughput nature of next-generation sequencers. However, the resulting graph may contain erroneous nodes or structures that were overlooked during the first phase. Hence, these erroneous structures must be removed or resolved, in the so called graph simplification phase, before the construction of the contigs. Following the graph simplification phase, the contigs are produced by finding the paths on the graph that connect the reads together. Subsequently, the scaffolding phase involves the filtering of the contigs, the detection of misassembled contigs and uncovering the relationships between them to build scaffolds [6]. Finally, the assembly assessment phase evaluates the assembled contigs/scaffolds in accordance with different metrics that reflect the quality, consistency, and accuracy of the algorithm used in the reconstruction process [7, 8].

There are many differing viewpoints when designing an assembler. Some designers rely on the early correction of errors in order to facilitate the remaining phases of the assembly process (i.e., graph building and simplification) [9–15]. Other designers propose to delay the error correction phase to the graph simplification process since both these phases aim at removing errors. Moreover, merging these two phases would reduce the overall computation time [16–22]. Hence, there are stand-alone error correction tools, scaffolding tools, and assessment tools that perform these phases independently from the other assembly phases. Certain designers rely on these independent tools to complete the missing parts in their assemblers.

8.2.1 Error Correction Phase

Correcting the errors that result from sequencing platforms represents one of the major challenges in the next-generation environment. These errors vary from the presence of simple ambiguous bases to the occurrence of substitution and indel errors (see Chap. 4). By detecting these errors early, the assembly process can be more efficient during the latter stages. The general approach followed by most error correction algorithms is examining the richness of the reads (i.e., read coverage) produced by the next-generation sequencers as a key to distinguish between correct and incorrect reads. This approach can be disrupted by repeats and non-uniform sampling of genomic sequences, which can lead to ambiguous choices during error correction [23].

8.2.2 Graph Construction Phase

There are diverse paradigms for graph construction in accordance with different graph models. These paradigms must overcome a host of computational challenges in relation to graph representation and path-finding algorithms for the contigs building (algorithms and challenges are discussed in detail in Chap. 9). Paradigms can generally be categorized into four main categories: overlap-based construction, *k*-mers-based construction, greedy-based construction, and hybrid-based construction [24, 25]. Each of these paradigms and their accompanying challenges are discussed in more detail in Chap. 9 as well.

8.2.3 Graph Simplification Phase

As mentioned previously, some errors are not recognized during the error correction phase and can subsequently complicate the efforts of path-finding algorithms that attempt to connect reads and assemble accurate contigs. These errors form diverse structures in the assembly graph which must be filtered through identification and correction before the building of contigs is initiated.

8.2.4 Scaffolding Phase

The process of creating scaffolds is not as simple as the process of creating contigs. The goal of the scaffolding process is to order and orient contigs that result from the assembly process. The scaffolding process is guided by paired-end reads that filter contigs, detect misassembled ones, and allow accurate contig extension into the repeated regions [6, 26].

8.2.5 Assembly Assessment Phase

Assessing the performance of an assembler is dependent on the metric(s) used during the evaluation process. One of these approaches targets the contiguity of the resulting contigs/scaffolds and utilizes different statistical metrics to assess the final assembled sequence [27–34]. Another approach scrutinizes the accuracy of the assembled contigs/scaffolds and uses one of the previously finished genomes as a reference to assess the draft sequence [29, 31]. Additional evaluative strategies include examining the constraints imposed by paired-end libraries, the nature of the sequences being assembled and the sequencing experiments themselves [31, 35, 36].

Since the assembler is a software program with a set of functionalities, it must be assessed not only in terms of its output but also in relation to other factors. These include responsiveness to user commands, the friendliness of the user interface components, and setup requirements. The evaluation of such functionalities allows the targeted assessment of the usability features of an assembler [37–39].

References

1. Pop M (2009) Genome assembly reborn: recent computational challenges. Briefings in bioinformatics 10 (4):354-366. doi:10.1093/bib/bbp026
2. Alkan C, Sajjadian S, Eichler EE (2011) Limitations of next-generation genome sequence assembly. Nat Methods 8 (1):61-65. doi:10.1038/nmeth.1527
3. Nagarajan N, Pop M (2013) Sequence assembly demystified. Nat Rev Genet 14 (3):157-167. doi:10.1038/nrg3367
4. Miller JR, Koren S, Sutton G (2010) Assembly algorithms for next-generation sequencing data. Genomics 95 (6):315-327. doi:10.1016/j.ygeno.2010.03.001
5. Pop M, Phillippy A, Delcher AL, Salzberg SL (2004) Comparative genome assembly. Briefings in bioinformatics 5 (3):237-248
6. El-Metwally S, Hamza T, Zakaria M, Helmy M (2013) Next-generation sequence assembly: four stages of data processing and computational challenges. PLoS Comput Biol 9 (12):e1003345. doi:10.1371/journal.pcbi.1003345
7. Earl D, Bradnam K, St John J, Darling A, Lin D et al. (2011) Assemblathon 1: a competitive assessment of de novo short read assembly methods. Genome research 21 (12):2224-2241. doi:10.1101/gr.126599.111
8. Bradnam KR, Fass JN, Alexandrov A, Baranay P, Bechner M et al. (2013) Assemblathon 2: evaluating de novo methods of genome assembly in three vertebrate species. Gigascience 2 (1):10. doi:2047-217X-2-10
9. Ilie L, Fazayeli F, Ilie S (2011) HiTEC: accurate error correction in high-throughput sequencing data. Bioinformatics 27 (3):295-302. doi:10.1093/bioinformatics/btq653
10. Kao WC, Chan AH, Song YS (2011) ECHO: a reference-free short-read error correction algorithm. Genome research 21 (7):1181-1192. doi:10.1101/gr.111351.110
11. Kelley DR, Schatz MC, Salzberg SL (2010) Quake: quality-aware detection and correction of sequencing errors. Genome Biol 11 (11):R116. doi:10.1186/gb-2010-11-11-r116
12. Medvedev P, Scott E, Kakaradov B, Pevzner P (2011) Error correction of high-throughput sequencing datasets with non-uniform coverage. Bioinformatics 27 (13):i137-i141. doi:10.1093/bioinformatics/btr208
13. Salmela L, Schroder J (2011) Correcting errors in short reads by multiple alignments. Bioinformatics 27 (11):1455-1461. doi:10.1093/bioinformatics/btr170
14. Schroder J, Schroder H, Puglisi SJ, Sinha R, Schmidt B (2009) SHREC: a short-read error correction method. Bioinformatics 25 (17):2157-2163. doi:10.1093/bioinformatics/btp379
15. Yang X, Dorman KS, Aluru S (2010) Reptile: representative tiling for short read error correction. Bioinformatics 26 (20):2526-2533. doi:10.1093/bioinformatics/btq468
16. Boetzer M, Henkel CV, Jansen HJ, Butler D, Pirovano W (2011) Scaffolding pre-assembled contigs using SSPACE. Bioinformatics 24 (4):578-579
17. Dayarian A, Michael TP, Sengupta AM (2010) SOPRA: Scaffolding algorithm for paired reads via statistical optimization. BMC bioinformatics 11:345. doi:10.1186/1471-2105-11-345
18. Donmez N, Brudno M (2013) SCARPA: scaffolding reads with practical algorithms. Bioinformatics 29 (4):428-434. doi:10.1093/bioinformatics/bts716
19. Gao S, Sung WK, Nagarajan N (2011) Opera: reconstructing optimal genomic scaffolds with high-throughput paired-end sequences. J Comput Biol 18 (11):1681-1691. doi:10.1089/cmb.2011.0170
20. Gritsenko AA, Nijkamp JF, Reinders MJ, de Ridder D (2012) GRASS: a generic algorithm for scaffolding next-generation sequencing assemblies. Bioinformatics 28 (11):1429-1437. doi:10.1093/bioinformatics/bts175
21. Koren S, Treangen TJ, Pop M (2011) Bambus 2: scaffolding metagenomes. Bioinformatics 27 (21):2964-2971. doi:10.1093/bioinformatics/btr520
22. Salmela L, Makinen V, Valimaki N, Ylinen J, Ukkonen E (2011) Fast scaffolding with small independent mixed integer programs. Bioinformatics 27 (23):3259-3265. doi:10.1093/bioinformatics/btr562

23. Yang X, Chockalingam SP, Aluru S (2013) A survey of error-correction methods for next-generation sequencing. Briefings in bioinformatics 14 (1):56-66. doi:10.1093/bib/bbs015
24. Medvedev P, Brudno M (2009) Maximum likelihood genome assembly. J Comput Biol 16 (8):1101-1116. doi:10.1089/cmb.2009.0047
25. Medvedev P, Georgiou K, Myers G, Brudno M (2007) Computability of Models for Sequence Assembly. In: Giancarlo R, Hannenhalli S (eds) Algorithms in Bioinformatics, vol 4645. Lecture Notes in Computer Science. Springer Berlin Heidelberg, pp 289-301. doi:10.1007/978-3-540-74126-8_27
26. Chaisson MJ, Brinza D, Pevzner PA (2009) De novo fragment assembly with short mate-paired reads: Does the read length matter? Genome research 19 (2):336-346. doi:10.1101/gr.079053.108
27. Church DM, Goodstadt L, Hillier LW, Zody MC, Goldstein S et al. (2009) Lineage-specific biology revealed by a finished genome assembly of the mouse. PLoS Biol 7 (5):e1000112. doi:10.1371/journal.pbio.1000112
28. Colbourne JK, Pfrender ME, Gilbert D, Thomas WK, Tucker A et al. (2011) The ecoresponsive genome of Daphnia pulex. Science 331 (6017):555-561. doi:10.1126/science.1197761
29. Li R, Fan W, Tian G, Zhu H, He L et al. (2010) The sequence and de novo assembly of the giant panda genome. Nature 463 (7279):311-317. doi:10.1038/nature08696
30. Lin Y, Li J, Shen H, Zhang L, Papasian CJ et al. (2011) Comparative studies of de novo assembly tools for next-generation sequencing technologies. Bioinformatics 27 (15):2031-2037. doi:10.1093/bioinformatics/btr319
31. Lindblad-Toh K, Wade CM, Mikkelsen TS, Karlsson EK, Jaffe DB et al. (2005) Genome sequence, comparative analysis and haplotype structure of the domestic dog. Nature 438 (7069):803-819. doi:nature04338
32. Liu Y, Qin X, Song XZ, Jiang H, Shen Y et al. (2009) Bos taurus genome assembly. BMC genomics 10:180. doi:10.1186/1471-2164-10-180
33. Locke DP, Hillier LW, Warren WC, Worley KC, Nazareth LV et al. (2011) Comparative and demographic analysis of orang-utan genomes. Nature 469 (7331):529-533. doi:10.1038/nature09687
34. Ming R, Hou S, Feng Y, Yu Q, Dionne-Laporte A et al. (2008) The draft genome of the transgenic tropical fruit tree papaya (Carica papaya Linnaeus). Nature 452 (7190):991-996. doi:10.1038/nature06856
35. Huson DH, Halpern AL, Lai Z, Myers EW, Reinert K et al. Comparing Assemblies Using Fragments and Mate-Pairs. In: WABI '01 Proceedings of the First International Workshop on Algorithms in Bioinformatics Århus, Denmark, 2001. Springer Berlin Heidelberg, pp 294-306
36. Phillippy AM, Schatz MC, Pop M (2008) Genome assembly forensics: finding the elusive misassembly. Genome Biol 9 (3):R55. doi:10.1186/gb-2008-9-3-r55
37. Golovko G, Khanipov K, Rojas M, Martinez-Alcantara A, Howard JJ et al. (2012) Slim-Filter: an interactive windows-based application for illumina genome analyzer data assessment and manipulation. BMC bioinformatics 13:166. doi:10.1186/1471-2105-13-166
38. Powell DR, Seemann T (2013) VAGUE: a graphical user interface for the Velvet assembler. Bioinformatics 29 (2):264-265. doi:10.1093/bioinformatics/bts664
39. Zhang W, Chen J, Yang Y, Tang Y, Shang J et al. (2011) A practical comparison of de novo genome assembly software tools for next-generation sequencing technologies. PLoS One 6 (3):e17915. doi:10.1371/journal.pone.0017915

Chapter 9
Approaches and Challenges
of Next-Generation Sequence Assembly Stages

Abstract The process of sequence assembly in the next-generation environment is broken down into five stages. We introduced all these stages in Chap. 8. Here, we will discuss four of these stages in detail and present the different approaches followed in each of them. Additionally, we will debate the challenges that face each stage and their stage-specific implementation approaches. The fifth stage, the assessment of the assembly, will be discussed separately in Chap. 10.

9.1 Introduction

Resolving the sequence assembly problem in the next-generation environment has involved a number of diverse approaches and faces many difficult challenges. These challenges can be viewed from the analogy of solving a jigsaw puzzle where increasing the number of pieces requires a strengthened effort to develop a solution. In sequence assembly, the "effort" is represented by the computational resources required for the assembly. Furthermore, the various pieces of the puzzle are not the same; some pieces have distinct features that present a direct indication of their probable location (e.g., the corners and edges) while other pieces may be similar in appearance resulting in ambiguity when attempting to identify their precise position. Moreover, trying to solve the puzzle without a picture of reference presents additional challenges when trying to bring the pieces together in a cohesive manner [1, 2]. From a sequence assembly perspective, these issues correspond to the process of assembling the high throughput short-reads that result from the next-generation sequencers. Potentially disturbed short-reads caused by sequencing errors as well as sequence repeats complicate the assembly process. Moreover, assembling these short-reads without a reference sequence (or de novo assembly) presents additional issues in comparison to the comparative assembly process [3, 4]. Here, we will discuss the various algorithms and challenges that are differentially involved in each phase of the sequence assembly process. Furthermore, we will provide examples of the different standalone tools that are implemented in each phase of the process.

S. El-Metwally et al., *Next Generation Sequencing Technologies*
and Challenges in Sequence Assembly, SpringerBriefs in Systems Biology 7,
DOI 10.1007/978-1-4939-0715-1_9, © The Authors 2014

9.2   Error Correction Phase

There are three approaches for correcting next-generation sequencing errors based on different algorithms and data structures that have been discussed in Chap. 2; these include k-mer-based error correction, suffix tree/array-based error correction, and alignment-based error correction. Additionally, some tools follow a hybrid approach where they combine two or more of these methodologies in order to obtain better error correction results. Since the implementation of each of these three approaches varies from one tool to another, we will discuss each of them with specific examples of their differential levels of implementation.

9.2.1   K-mer-Based Error Correction

One of these types of approaches is based on extracting k-mers from the reads, calculating their scores, and using the high-scoring k-mers to correct the low-scoring ones. The scores of k-mers are calculated according to different factors such as their frequencies and the quality of their bases. This idea has been implemented in different standalone error correction tools such as Quake [5]. Quake utilizes base quality values to compute the weights of k-mers. Each k-mer possesses a weight value which is computed as a weighted sum of all its base quality scores. Subsequently, Quake chooses a cutoff point (M) to differentiate between trusted and untrusted k-mers. This is accomplished by modeling the weight histogram as a combination of Gaussian and Zeta probability distributions for trusted k-mers, and Gamma probability distributions for untrusted k-mers. Untrusted k-mers in reads are located heuristically, using low-weighted values as an indicator, and are corrected greedily until all reads become error-free.

Reptile [6] is another standalone tool for error correction, and is based on a slightly different concept. Instead of decomposing reads into a set of k-mers and computing the frequency of each k-mer, Reptile decomposes reads into tiles (two or more overlapping and non-overlapping k-mers) and computes the frequency of each tile. Using tiles instead of k-mers helps to retain the k-mer context, which represents another factor that improves the quality of error correction methods. Each tile has a weight value, which is calculated using the quality score values and the context of k-mers. If this value is lower than a specified threshold, the tile is considered an erroneous tile. To correct erroneous tiles, Reptile builds a hamming graph where each node represents a tile and the edges represent the number of different characters between the tiles (hamming distance). A possible corrective solution for an erroneous node is detected if an edge connects this node to another with a hamming distance $\leq$ threshold, allowing the node to be converted into another.

Hammer [7] implements the same idea as Reptile's hamming graph but from a different perspective through the utilization of a spaced seed. Hammer clusters

k-mers according to their similarities and defines a consensus (a high-frequency error-free k-mer) for each cluster. This reduces the search space into subspaces with each subspace having its own set of k-mers. Furthermore, this idea can parallelize the subspace search, which helps to reduce the runtime.

9.2.2 Suffix Tree/Array-Based Error Correction

Another approach for correcting next-generation sequencing errors is based on using the suffix tree/array (see Chap. 2) as a data structure to organize the variable-size k-mers with their associated scores. This approach has been implemented in some software packages like SHREC [8], which is based on the suffix tree approach. It stores all of the suffixes of the reads in a weighted suffix tree, where the frequency of each suffix is included in each internal node. Using a defined formula based on the number, length, and suffix size of the reads, SHREC computes the expected frequencies of each suffix in the set of reads. Subsequently, the actual frequency of each suffix in the set of reads is computed. If the expected values differ from the actual computed ones, the substrings that belong to these values are detected as containing errors.

HiTEC [9] utilizes a suffix array data structure rather than a suffix tree. It starts to correct erroneous positions in the reads by extracting the set of suffixes from them. Next, it defines a cluster for each suffix, which contains all of the instances of that suffix followed by different characters. All of the suffixes in one cluster are consecutive, allowing the construction of the suffix array as well as the computation of the longest common prefix (LCP) to be simpler. For each suffix in the cluster, the supported value is computed. A suffix with a highly supported value is chosen as a candidate solution to correct those with similar low support values.

9.2.3 Alignment-Based Error Correction

The third approach of error correction uses alignment algorithms to detect erroneous bases centered on aligning high trusted reads (i.e., according to their k-mer frequencies) to low trusted ones. This approach has been implemented in different standalone tools such as Coral [10], which stores a hash table of k-mers and their associated reads in forward and backward directions. Coral aligns the reads by selecting each read as a base and searching for a set of reads which share at least one common k-mer with the base read. These sets of reads represent the k-mer neighbor list of the base read. Subsequently, Coral chooses one read from the k-mer neighbor list and aligns it with a base read using a variant of the Needleman–Wunsch algorithm [11]. The resulting consensus is used again to align the remaining reads until the reads in the neighbor list are all aligned with the base read. The alignment procedure is used as the basis to correct errors and indicate indels and substitution points in the reads.

ECHO [12] is another error correction filter based on the alignment approach. This filter has two major stages, neighbor finding and maximum posteriori error correction. Unlike k-mer-based error correction methods and their calculated frequencies, ECHO relies on finding the overlap between reads, before setting the parameters automatically and estimating the error characteristics in each specific sequencing run using an expectation-maximization (EM) procedure. ECHO corrects substitution errors in the reads using a maximum posteriori procedure, which relies on the construction of a position-dependent matrix that records the substitution error rate at position i between any two bases in the set {A, G, C, T} [13]. For each overlapping position, the consensus base is determined as the one with the maximum posteriori estimation using the quality scores of the bases and the overlap information.

Since the next-generation sequencing technologies are characterized with their high throughput short-reads distorted by different levels of sequencing errors and genomic repeats, detecting and correcting these errors early plays a crucial role in the success and quality of the assembly process. There are many challenges encountered during the error correction phase. These include choosing a set of suitable parameters according to the complexity of the data sets being corrected (e.g., genome, transcriptome, and metagenome) and their different characteristics (e.g., different levels of errors, read coverage, and length), differentiating single nucleotide polymorphisms (SNP) from sequencing errors, improving the use of paired-end reads to detect and resolve genomic repeats early, and increasing the performance of error correcting algorithms in terms of their CPU time and memory to overcome the continuous deluge of data resulting from high throughput sequencers [14].

Tables 9.1 and 9.2 summarize the technical and practical details of some standalone error-correction tools.

9.3 Graph Construction Phase

The filtered reads that result from the error correction phase are the input for the graph construction phase that represents the second phase of the next-generation sequence assembly process. In this phase, the reads and the relationship between them (e.g., the overlap) are formulated as a graph of nodes where the sequence reads are represented by the nodes and their relationships by the edges. This representation was mainly selected to overcome the limitation of available computational resources that are necessary to process the high throughput nature of next-generation sequencers. Among the next-generation assembly tools and pipelines, three main paradigms are generally recognized in the construction of the graphs; these are overlap-based construction, k-mers-based construction, and greedy-based

Table 9.1 Technical comparison of in standalone error correction programs

Error Correction tools	Latest version	Operating system	Programming language	Single PC/cluster	Open source	Websites	Reference
Quake[a]	(V0.3)	Linux (64 bits)	C++, Python	Single	Y	http://www.cbcb.umd.edu/software/quake	[5]
Reptile[a]	(V1.1)	Linux (64 bits)	C++, Perl	Single	Y	http://aluru-sun.ece.iastate.edu/doku.php?id=reptile	[6]
Hammer[a]	(V0.2)	Linux (64 bits)	C++, Perl	Single	Y	http://bix.ucsd.edu/projects/hammer	[7]
Musket[a]	(V1.1)	Linux	C++	Shared-memory computers	Y	http://musket.sourceforge.net	[60]
SHREC	(V2.2)	Linux (64 bits)/Mac OS X/Windows	Java	Single/cluster	Y	http://sourceforge.net/projects/shrec-ec	[8]
HiTEC[a]	(V1.0.2)	Linux (64 bits)	C++	Single	Y	http://www.csd.uwo.ca/~ilie/HiTEC	[9]
Coral[a]	(V1.4)	Linux	C++	Single	Y	http://www.cs.helsinki.fi/u/lmsalmel/coral	[10]
ECHO[a]		Linux/Mac OS X/Windows	C++, Python	Single	Y	http://uc-echo.sourceforge.net	[12]
Hybrid-SHREC		Linux (64 bits)	Java	Single	Y	http://www.cs.helsinki.fi/u/lmsalmel/hybrid-shrec/	[61]
PBcR[a]		Linux (64 bits)	C, C++, Perl	Single/cluster	Y	http://sourceforge.net/apps/mediawiki/wgs-assembler/	[62]

[a] Personal communications with authors

Table 9.2 Practical comparison of in standalone error correction programs

Tool	Correcting approach	Targeted errors	Sequencing platform	Input file formats	Output file formats
Quake[a]	K-mer	Substitution	Illumina	fastq	fastq
Reptile[a]	K-mer	Substitution	Illumina	fastq	.fa, .errors[b]
Hammer[a]	K-mer	Substitution	Illumina	fastq	raw k-mers
Musket[a]	K-mer	Substitution	Illumina	fasta/fastq	fasta/fastq
SHREC	Suffix tree	Substitution	Illumina	fastq[S]	fastq[S] [8]
HiTEC[a]	Suffix array	Substitution	Illumina	fasta/fastq	fasta
Coral[a]	Alignment	Substitution, insertion/ deletion	Any platform	fasta/fastq	fasta/fastq
ECHO[a]	Alignment	Substitution	Illumina	.txt	fastq
Hybrid-SHREC	Hybrid	Substitution, insertion/ deletion	Any platform	fasta[c]	fasta
PBcR[a]	Hybrid	Substitution, insertion/ deletion	PacBio RS 454 Illumina	fastq[d] SFF fasta	FRG fasta qual fastq

[a] Personal communications with authors
[b] Files recording error positions and bases can be converted into fasta/fastq
[c] Supported base or color space
[d] Illumina and PacBio RS formats, also there are tools for converting fasta files to fastq-compatible files
[S] Speculated, based on sequencing platforms

construction [15, 16]. Furthermore, a hybrid construction approach that combines two or more of these approaches is implemented in some tools [15, 16]. Here, we will discuss each of the three approaches as well as some of the challenges that face them.

9.3.1   Overlap-Based Graph Construction Approach

This approach is also known as Overlap-Layout-Consensus (OLC) and aims to find Hamiltonian cycles (see Chap. 2). There are three steps in this approach; (1) an overlap step that computes all pairwise alignments among reads to detect overlaps, (2) a layout step that is responsible for finding Hamiltonian cycles in the graph (these cycles correspond to created contigs from the set of reads being assembled), and (3) a consensus step that combines overlaps among the set of reads in the assembly set.

The performance of the tools that implement this paradigm depends on the length of reads, since the overlap detection can be more effective in long reads (e.g., reads produced by Roche 454 or Pacific Biosciences systems) rather than shorter reads (e.g., reads produced by Illumina or SOLiD sequencers) [1, 17]. Furthermore, the length of the overlap is another challenge that raises the question of how many characters are sufficient to detect overlaps among reads. Moreover, there are no efficient algorithms to find the optimal path through an assembly graph using all set of reads since the assembly problem is reported theoretically as NP-hard [16]. According to this classification, different heuristics algorithms are used to find an approximation of the optimal path with the most interesting one being the greedy algorithm (see Chap. 2). The greedy algorithm connects to the reads with maximum overlap length first and then proceeds to the other reads [17–22].

Another form of the overlap graph is the string graph [23] which overcomes previous concerns of overlap computation phase complexity during the classical OLC approach. By introducing FM-index, a string indexing structure, the set of overlaps among reads can be computed roughly in linear time, which improves the speed and performance of string-based assemblers for short-reads sequence assembly [24–26].

9.3.2 K-mer-Based Graph Construction Approach

This paradigm is also known as the De Bruijn graph and aims to find Eulerian cycles (see Chap. 2). This approach is met by many challenges, including finding relationships among the set of k-mers rather than reads which may result in information loss in the context of k-mers. Additionally, an increase in hardware requirements in relation to the accurate processing and storage of these k-mers may also be required. This paradigm also needs continuous refinement in order to manage high-coverage k-mers with high-error profiles. Moreover, this form of graph representation is still sensitive to the k parameter that must be chosen accurately to increase the chance of overlap among true k-mers and decrease the chance of false overlaps [1, 27–34].

A number of studies have addressed the issue of overcoming the large amount of memory needed to store a graph of k-mers. Novel solutions to reduce memory requirements have included storing a subset of k-mers in memory rather than the entire set. In this case, the select group of k-mers being stored represented a $1/g$ subsample of different k-mers in the set of reads being assembled [35]. Additionally, Conway et al. proposed a concise bit map structure to represent a De Bruijn graph [36], Bowe et al. used an extension of the Burrows–Wheeler transform to efficiently index and compress the set of graph nodes and edges [37], while Chikhi and Rizk used the Bloom filter to represent the De Bruijn graph efficiently in memory [38]. Furthermore, Salikhov et al. improved graph representation in memory by using a combination of cascading Bloom filters [39].

Another approach involves the paired De Bruijn graph, which incorporates paired-end reads early in the graph construction phase rather than incorporating them in the later stages during the scaffolding process. Paired-end reads have had a great impact on resolving most of the challenges facing current assemblers, including genomic repeats and misassembled contigs. By employing a paired De Bruijn graph instead of the traditional version in the k-mer based assembler, the information contained in paired-end reads can be utilized in each phase of the assembly process rather than only in the later scaffolding phase [40]. Moreover, a rectangle version of the De Bruijn graph was recently introduced to model the assembly problem in the two-dimensional space of the assembly process and to utilize paired-end reads as substrings of the originally reconstructed string [2].

9.3.3   Greedy-Based Graph Construction Approach

The main idea behind this approach is to utilize the greedy algorithm that is described in Chap. 2 and the previous section. In this case, the greedy algorithm is used as a path-finding algorithm during sequence assembly problems. This approach is suitable for small-sized genomes but suffers from the non-optimal nature of the greedy algorithm which may resort to local solutions while tackling assembly problems [41–44].

9.4   Graph Simplification Phase

Approaches to graph simplification can vary from the simple removal of low coverage nodes and their associated arcs to the correction of more complex erroneous structures on the graph [45]. One of these erroneous structures, which is a result of oversampling the sequencing technology, is transitive edges [23]. The edge $E_a \rightarrow E_b$ is considered transitive if the graph contains this edge with another one $E_a \rightarrow E_c \rightarrow E_b$ and hence this transitive edge is included implicitly in the second edge. The graph simplification phase (or graph simplifier) resolves this error by removing transitive edges iteratively and hence reducing the graph complexity by a factor of $c = NL/G$ where c is the oversampling rate, N is the number of reads, L is the length of reads, and G represents the genome size [18, 46, 47].

Other forms of erroneous structures include dead ends (also known as tips) that result from the low-depths of k-mer coverage or sequencing errors that have produced a mixture of correct and incorrect k-mers. The graph simplifier resolves dead ends by removing them from the graph. The removal process is based on testing the depth of all paths on the graph according to the specified minimum depth threshold, k-mers coverage or based on the length of the k parameter [18, 21, 27, 28, 31, 34].

Bubbles or bulges are erroneous structures that result from inexact repeat sequences. Their location is detected on the graph by tracing divergent paths that converge after the k points. The k parameter is a specified threshold that is set according to the assembly strategy. Some assemblers do not use the k-parameter to detect bubbles, instead tracing the graph forwards after each diverging point and then stopping when the convergence point is reached. All paths between those two points are filtered according to their k-mers coverage and quality scores, or aligned together to determine their shared consensus sequences [31, 34, 48].

One of the most important challenges that face current assemblers in the next-generation environment is repeats. Repeats allow more than one possible construction of the target sequences. The length of the repeats greatly affects the process of repairing them. Repeats possessing a length shorter than the read length and a graph structure of nodes with equal incoming and outgoing edges N are simplified by removing the repeated nodes from the graph and expanding its edges into N parallel paths. These kind of repeats are called X-cuts or tangles [28, 30, 31]. On the other hand, countering repeats which have a length longer than the read length require different heuristics techniques based on paired-end constraints [34, 49].

Another form of simplification involves a reduction in the large number of nodes in assembly graphs to decrease memory requirements and reduce costs. For each node Q that has one outgoing edge to another node G that has only one incoming edge, the two consecutive nodes will be merged in the simplified graph to represent one node. This simplification process corresponds to the concatenation of two character strings [21, 22, 30, 31].

9.5 Scaffolding Phase

After the graph simplification phase is completed, the corrected graph is traversed to build long contigs. These contigs are linked together in a further step to form scaffolds. By mapping paired-end reads to the set of contigs being joined, the correct contigs are linked together based on the position of the reads, their known orientation, and the insert size. This information can also be used to detect chimeric contigs that result from misassembled contigs from two different genomic locations. Moreover, the frequency of paired-end reads can be used as a criterion to support the link between two contigs [49].

There are many challenges that face the scaffolding process including disturbances due to sequencing errors that are not detected during the early error correction or graph simplification phases. These errors can form erroneous structures in the contig-connectivity graph as well as the short-read assembly graph. Furthermore, misassembled and chimeric contigs represent erroneous nodes that can violate the constraints imposed by paired-end libraries. In addition, distortions in paired-end libraries that can result from sequencing experiments and chimeric paired-end reads represent erroneous links on the graph. These links must be detected and removed because they interrupt the solution of any scaffolding algorithm. Resolving repeat structures on the contig-connectivity graph and the mapping of paired-end reads to several locations in the set of contigs impose additional challenges on the scaffolding modules [50–57].

The scaffolder is an independent tool or module in the assembly software that takes the set of contigs and the set of paired-end reads as input in order to connect them together to build longer sequence stretches (e.g., whole genome or chromosome). In this case, the orientation of paired-end reads and the separation distance between them is approximately known. The goal of any scaffolding algorithm is to take a majority of voting from a large number of paired-end reads to minimize inconsistencies that result from misassembled contigs. While achieving this goal is NP-hard, there are different heuristics that can approximate the solution to the scaffolding problem [58].

Some assemblers use the previously created De Bruijn graph to build scaffolds after incorporating paired-end reads into the graph by aligning them against contig paths, or alternatively using heuristic approaches to locate them directly on these paths [27, 34, 59]. Other assemblers use the contig-connectivity graph to build scaffolds by assigning a node to each contig. In this regard, two contigs are linked together if they are satisfied with the constraints encoded by the paired-end reads. This graph may contain repeated nodes (contigs) as well as transitive, associative, and erroneous connections that need repairing before the scaffolding process begins. The greedy approach is usually used to find the scaffold paths on the graph through maximizing the number of supporting paired-end constraints [56] or visiting contigs in order to increase their lengths [50].

Some assemblers produce the assembled sequences as a set of contigs while others may generate them as a set of scaffolds via their own scaffolding modules such as *Euler-SR*, *Velvet*, *ALLPATHS-LG*, *SOAPdenovo*, and *Celera assembler* [47].

9.5.1   Scaffolders

The scaffolding process can be performed by standalone scaffolders independently from the other assembly phases. One of these standalone scaffolders is Bumbus [56], which was originally designated for Sanger sequence reads. A newer version of this scaffolder is called Bumbus2 [55] and is designed for next-generation metagenomic sequences. Generally, Bumbus involves three steps in building scaffolds: (1) determining the orientation of the contigs, (2) assessing their position, and (3) repairing the contig graph by reducing the graph size, removing erroneous edges, and resolving conflict repeats. Bumbus and its successor build scaffolds based on creating contig-connectivity graphs. They collect the links between each pair of contigs based on paired-end reads and take the majority among these links to set the contig orientation. Subsequently, the distance constraints are checked among each pair of reads to determine valid and invalid links among the contigs. Two contigs are linked together if they share a maximum number of valid links. Additionally, the overlapping sequences among two contigs are used to link contigs in the graph.

SCARPA [52] is another scaffolder that begins its pipeline with preprocessing paired-end reads, checking their contradictory links corresponding to assembled contigs and re-computing the insert size distribution of each paired-end library. Based on satisfying a maximum number of paired-end constraints, SCARPA assigns an orientation for each contig and discards erroneous and contradictory links as well as misassembled contigs from the contig graph. By using the linear programming model, SCARPA provides an exact order for each contig within a scaffold [52].

SSPACE [50] starts its scaffolding process by removing paired-end reads that contain non-valid DNA characters and then aligns them against assembled contigs using the Bowtie aligner. The orientation and position associated to each pair is hashed so that they can be retrieved easily. SSPACE connects two contigs based on satisfying the maximum number of constraints imposed by their mapped paired-end reads. Each scaffold is constructed greedily by combining contigs that have large lengths first and share k supported links between them. The correct order of each contig within each scaffold is determined by the library insert size imposed by their mapped paired-end reads or specified threshold in the case of multiple order choices.

SOPRA [51] is another scaffolding module that can be easily integrated with any of the existing assemblers in the next-generation environment. It formulates the scaffolding process as an optimization problem by statistically linking contigs based on satisfying a maximum number of paired-end constraints. The correct orientation and position of each contig is determined by the same approach utilized during the removal of links or contigs that violate these constraints. By removing erroneous nodes, the scaffolding graph can be partitioned into separate components. These components can be solved independently by optimizing the sets of paired-end constraints. Moreover, SOPRA can manage the color-space data produced by the SOLiD sequencer.

Opera [53] partitions the contigs graph from a different perspective based on graph contraction. By using a graph bandwidth formulation, Opera solves the scaffolding problem with a fixed parameter tractable algorithm.

MIP Scaffolder [57] divides the contig-connectivity graph into subgraphs and tries to find the local solution to each one using mixed integer programming. The final scaffolding solution comes from integrating the information contained in these local solutions into the global one and removes unnecessary assembled contigs. Unlike other scaffolders, the size of each subgraph is restricted to exhibit accurate production of scaffolds.

GRASS [54] is another scaffolder that uses the concept of mixed integer programming for creating scaffolds. It designs an objective function based on the orientation, orders, and distances of the contigs and tries to optimize this function by satisfying a maximum number of paired-end constraints using an expectation-maximization approach.

Table 9.3 summarizes the technical and practical details of some standalone scaffolding tools.

Table 9.3 Practical and technical challenges in standalone scaffolders

Scaffolders	Operating system	Programming language	Single PC/cluster	Open source	Paired-end libraries	Input file formats	Output file formats	Websites	Reference
Bumbus2[a] (V2.0)	Linux, Sun/Solaris Alpha/Ultrix Darwin OS X	C++, Python, Perl	Single	Y	Illumina 454	AMOS[c]	fasta, agp, dot	http://amos.sf.net	[55]
SSPACE[a]	Linux	Perl	Single	Y (Basic v) N (Premium)	Illumina 454 SOLiD	fasta/fastq	fasta	www.baseclear.com/ bioinformatics-tools/	[50]
SOPRA (V1.4.6)	Linux (64 bits) Mac OS X	Perl	Single	Y	Illumina SOLiD[b]	fasta, sam[b]	fasta[b]	http://www.physics.rutgers.edu/ ~anirvans/SOPRA/	[51]
Opera[a] (V1.3.1)	Linux	Java/C++	Single	Y	Any platform	fasta,[d] sam/ bowtie	fasta [8]	http://sourceforge.net/ projects/operasf	[53]
MIP Scaffolder[a] (V0.5)	Linux (64 bits)	C++, Perl	Single	Y	Illumina	fasta, sam	fasta	http://www.cs.helsinki.fi/u/ lrrsalmel/mip-scaffolder/	[57]
SCARPA[a]	Linux (64 bits)	C++, Perl	Single	Y	Illumina SOLiD	fasta, sam	fasta	http://compbio.cs.toronto.edu/ hapsembler/scarpa.html	[52]
GRASS[a]	Linux	C++	Single	Y	Illumina	fasta, fastq[e]	fasta	http://code.google.com/ p/tud-scaffolding/	[54]

[a] Personal communications with authors

[b] Users experiences and communities websites

[c] There are utilities available to import a variety of different data formats into AMOS

[d] For preprocessing, read files can be in fasta/fastq

[e] Fasta (contigs and reference genomes) and fastq (reads)

References

1. Pevzner PA, Tang H, Waterman MS (2001) An Eulerian path approach to DNA fragment assembly. Proceedings of the National Academy of Sciences of the United States of America 98 (17):9748-9753. doi:10.1073/pnas.171285098
2. Vyahhi N, Pyshkin A, Pham S, Pevzner P (2012) From de Bruijn Graphs to Rectangle Graphs for Genome Assembly. In: Raphael B, Tang J (eds) Algorithms in Bioinformatics, vol 7534. Lecture Notes in Computer Science. Springer Berlin Heidelberg, pp 249-261. doi:10.1007/978-3-642-33122-0_20
3. Martin JA, Wang Z (2011) Next-generation transcriptome assembly. Nature reviews Genetics 12 (10):671-682. doi:10.1038/nrg3068
4. Pop M, Phillippy A, Delcher AL, Salzberg SL (2004) Comparative genome assembly. Briefings in bioinformatics 5 (3):237-248
5. Kelley DR, Schatz MC, Salzberg SL (2010) Quake: quality-aware detection and correction of sequencing errors. Genome Biol 11 (11):R116. doi:10.1186/gb-2010-11-11-r116
6. Yang X, Dorman KS, Aluru S (2010) Reptile: representative tiling for short read error correction. Bioinformatics 26 (20):2526-2533. doi:10.1093/bioinformatics/btq468
7. Medvedev P, Scott E, Kakaradov B, Pevzner P (2011) Error correction of high-throughput sequencing datasets with non-uniform coverage. Bioinformatics 27 (13):i137-i141. doi:10.1093/bioinformatics/btr208
8. Schroder J, Schroder H, Puglisi SJ, Sinha R, Schmidt B (2009) SHREC: a short-read error correction method. Bioinformatics 25 (17):2157-2163. doi:10.1093/bioinformatics/btp379
9. Ilie L, Fazayeli F, Ilie S (2011) HiTEC: accurate error correction in high-throughput sequencing data. Bioinformatics 27 (3):295-302. doi:10.1093/bioinformatics/btq653
10. Salmela L, Schroder J (2011) Correcting errors in short reads by multiple alignments. Bioinformatics 27 (11):1455-1461. doi:10.1093/bioinformatics/btr170
11. Needleman SB, Wunsch CD (1970) A general method applicable to the search for similarities in the amino acid sequence of two proteins. J Mol Biol 48 (3):443-453. doi:0022-2836(70)90057-4
12. Kao WC, Chan AH, Song YS (2011) ECHO: a reference-free short-read error correction algorithm. Genome research 21 (7):1181-1192. doi:10.1101/gr.111351.110
13. Zhang Q, Pell J, Canino-Koning R, Chuang Howe CA, Brown T (under review) These are not the k-mers you are looking for: efficient online k-mer counting using a probabilistic data structure. Preprint arXiv: 1309:2975. In review, PloS One
14. Yang X, Chockalingam SP, Aluru S (2013) A survey of error-correction methods for next-generation sequencing. Briefings in bioinformatics 14 (1):56-66. doi:10.1093/bib/bbs015
15. Medvedev P, Brudno M (2009) Maximum likelihood genome assembly. J Comput Biol 16 (8):1101-1116. doi:10.1089/cmb.2009.0047
16. Medvedev P, Georgiou K, Myers G, Brudno M (2007) Computability of Models for Sequence Assembly. In: Giancarlo R, Hannenhalli S (eds) Algorithms in Bioinformatics, vol 4645. Lecture Notes in Computer Science. Springer Berlin Heidelberg, pp 289-301. doi:10.1007/978-3-540-74126-8_27
17. DiGuistini S, Liao NY, Platt D, Robertson G, Seidel M et al. (2009) De novo genome sequence assembly of a filamentous fungus using Sanger, 454 and Illumina sequence data. Genome Biol 10 (9):R94. doi:10.1186/gb-2009-10-9-r94
18. Hernandez D, Francois P, Farinelli L, Osteras M, Schrenzel J (2008) De novo bacterial genome sequencing: Millions of very short reads assembled on a desktop computer. Genome research 18 (5):802-809. doi:10.1101/gr.072033.107
19. Hossain M, Azimi N, Skiena S (2009) Crystallizing short-read assemblies around seeds. BMC bioinformatics 10 (Suppl 1):S16. doi:10.1186/1471-2105-10-s1-s16
20. Margulies M, Egholm M, Altman WE, Attiya S, Bader JS et al. (2005) Genome sequencing in microfabricated high-density picolitre reactors. Nature 437 (7057):376-380. doi:nature03959
21. Miller JR, Delcher AL, Koren S, Venter E, Walenz BP et al. (2008) Aggressive assembly of pyrosequencing reads with mates. Bioinformatics 24 (24):2818-2824. doi:10.1093/bioinformatics/btn548

22. Myers EW, Sutton GG, Delcher AL, Dew IM, Fasulo DP et al. (2000) A whole-genome assembly of Drosophila. Science 287 (5461):2196-2204
23. Myers EW (2005) The fragment assembly string graph. Bioinformatics 21 Suppl 2:ii79-85. doi:21/suppl_2/ii79
24. Gonnella G, Kurtz S (2012) Readjoiner: a fast and memory efficient string graph-based sequence assembler. BMC bioinformatics 13:82. doi:10.1186/1471-2105-13-82
25. Simpson JT, Durbin R (2010) Efficient construction of an assembly string graph using the FM-index. Bioinformatics 26 (12):i367-373. doi:10.1093/bioinformatics/btq217
26. Simpson JT, Durbin R (2012) Efficient de novo assembly of large genomes using compressed data structures. Genome research 22 (3):549-556. doi:10.1101/gr.126953.111
27. Butler J, MacCallum I, Kleber M, Shlyakhter IA, Belmonte MK et al. (2008) ALLPATHS: de novo assembly of whole-genome shotgun microreads. Genome research 18 (5):810-820. doi:10.1101/gr.7337908
28. Chaisson M, Pevzner P, Tang H (2004) Fragment assembly with short reads. Bioinformatics 20 (13):2067-2074. doi:10.1093/bioinformatics/bth205
29. Chaisson MJ, Brinza D, Pevzner PA (2009) De novo fragment assembly with short mate-paired reads: Does the read length matter? Genome research 19 (2):336-346. doi:10.1101/gr.079053.108
30. Chaisson MJ, Pevzner PA (2008) Short read fragment assembly of bacterial genomes. Genome research 18 (2):324-330. doi:10.1101/gr.7088808
31. Li R, Zhu H, Ruan J, Qian W, Fang X et al. (2010) De novo assembly of human genomes with massively parallel short read sequencing. Genome research 20 (2):265-272. doi:10.1101/gr.097261.109
32. Maccallum I, Przybylski D, Gnerre S, Burton J, Shlyakhter I et al. (2009) ALLPATHS 2: small genomes assembled accurately and with high continuity from short paired reads. Genome Biol 10 (10):R103. doi:10.1186/gb-2009-10-10-r103
33. Simpson JT, Wong K, Jackman SD, Schein JE, Jones SJ et al. (2009) ABySS: a parallel assembler for short read sequence data. Genome research 19 (6):1117-1123. doi:10.1101/gr.089532.108
34. Zerbino DR, Birney E (2008) Velvet: Algorithms for de novo short read assembly using de Bruijn graphs. Genome research 18 (5):821-829. doi:10.1101/gr.074492.107
35. Ye C, Ma ZS, Cannon CH, Pop M, Yu DW (2012) Exploiting sparseness in de novo genome assembly. BMC bioinformatics 13 Suppl 6:S1. doi:10.1186/1471-2105-13-S6-S1
36. Conway TC, Bromage AJ (2011) Succinct data structures for assembling large genomes. Bioinformatics 27 (4):479-486. doi:10.1093/bioinformatics/btq697
37. Bowe A, Onodera T, Sadakane K, Shibuya T (2012) Succinct de Bruijn Graphs. In: Raphael B, Tang J (eds) Algorithms in Bioinformatics, vol 7534. Lecture Notes in Computer Science. Springer Berlin Heidelberg, pp 225-235. doi:10.1007/978-3-642-33122-0_18
38. Chikhi R, Rizk G (2012) Space-Efficient and Exact de Bruijn Graph Representation Based on a Bloom Filter. In: Raphael B, Tang J (eds) Algorithms in Bioinformatics, vol 7534. Lecture Notes in Computer Science. Springer Berlin Heidelberg, pp 236-248. doi:10.1007/978-3-642-33122-0_19
39. Salikhov K, Sacomoto G, Kucherov G (Submitted) Using cascading Bloom filters to improve the memory usage for de Brujin graphs.
40. Medvedev P, Pham S, Chaisson M, Tesler G, Pevzner P (2011) Paired de bruijn graphs: a novel approach for incorporating mate pair information into genome assemblers. J Comput Biol 18 (11):1625-1634. doi:10.1089/cmb.2011.0151
41. Bryant DW, Jr., Wong WK, Mockler TC (2009) QSRA: a quality-value guided de novo short read assembler. BMC bioinformatics 10:69. doi:10.1186/1471-2105-10-69
42. Dohm JC, Lottaz C, Borodina T, Himmelbauer H (2007) SHARCGS, a fast and highly accurate short-read assembly algorithm for de novo genomic sequencing. Genome Res 17 (11):1697-1706. doi:gr.6435207
43. Jeck WR, Reinhardt JA, Baltrus DA, Hickenbotham MT, Magrini V et al. (2007) Extending assembly of short DNA sequences to handle error. Bioinformatics 23 (21):2942-2944. doi:10.1093/bioinformatics/btm451

44. Warren RL, Sutton GG, Jones SJ, Holt RA (2007) Assembling millions of short DNA sequences using SSAKE. Bioinformatics 23 (4):500-501. doi: 10.1093/bioinformatics/btl629

45. Miller JR, Koren S, Sutton G (2010) Assembly algorithms for next-generation sequencing data. Genomics 95 (6):315-327. doi:10.1016/j.ygeno.2010.03.001

46. Schmidt B, Sinha R, Beresford-Smith B, Puglisi SJ (2009) A fast hybrid short read fragment assembly algorithm. Bioinformatics 25 (17):2279-2280. doi:10.1093/bioinformatics/btp374

47. El-Metwally S, Hamza T, Zakaria M, Helmy M (2013) Next-generation sequence assembly: four stages of data processing and computational challenges. PLoS Comput Biol 9 (12):e1003345. doi:10.1371/journal.pcbi.1003345

48. Gnerre S, Maccallum I, Przybylski D, Ribeiro FJ, Burton JN et al. (2011) High-quality draft assemblies of mammalian genomes from massively parallel sequence data. Proceedings of the National Academy of Sciences of the United States of America 108 (4):1513-1518. doi:10.1073/pnas.1017351108

49. Zerbino DR, McEwen GK, Margulies EH, Birney E (2009) Pebble and rock band: heuristic resolution of repeats and scaffolding in the velvet short-read de novo assembler. PLoS One 4 (12):e8407. doi:10.1371/journal.pone.0008407

50. Boetzer M, Henkel CV, Jansen HJ, Butler D, Pirovano W (2011) Scaffolding pre-assembled contigs using SSPACE. Bioinformatics 24 (4):578-579

51. Dayarian A, Michael TP, Sengupta AM (2010) SOPRA: Scaffolding algorithm for paired reads via statistical optimization. BMC bioinformatics 11:345. doi:10.1186/1471-2105-11-345

52. Donmez N, Brudno M (2013) SCARPA: scaffolding reads with practical algorithms. Bioinformatics 29 (4):428-434. doi:10.1093/bioinformatics/bts716

53. Gao S, Sung WK, Nagarajan N (2011) Opera: reconstructing optimal genomic scaffolds with high-throughput paired-end sequences. J Comput Biol 18 (11):1681-1691. doi:10.1089/cmb.2011.0170

54. Gritsenko AA, Nijkamp JF, Reinders MJ, de Ridder D (2012) GRASS: a generic algorithm for scaffolding next-generation sequencing assemblies. Bioinformatics 28 (11):1429-1437. doi:10.1093/bioinformatics/bts175

55. Koren S, Treangen TJ, Pop M (2011) Bambus 2: scaffolding metagenomes. Bioinformatics 27 (21):2964-2971. doi:10.1093/bioinformatics/btr520

56. Pop M, Kosack DS, Salzberg SL (2004) Hierarchical scaffolding with Bambus. Genome research 14 (1):149-159. doi:10.1101/gr.1536204

57. Salmela L, Makinen V, Valimaki N, Ylinen J, Ukkonen E (2011) Fast scaffolding with small independent mixed integer programs. Bioinformatics 27 (23):3259-3265. doi:10.1093/bioinformatics/btr562

58. Huson DH, Reinert K, Myers EW (2002) The greedy path-merging algorithm for contig scaffolding. Journal of the ACM 49 (5):603 - 615

59. Medvedev P, Brudno M (2008) Ab initio whole genome shotgun assembly with mated short reads. Paper presented at the Proceedings of the 12th annual international conference on Research in computational molecular biology, Singapore

60. Liu Y, Schroder J, Schmidt B (2013) Musket: a multistage k-mer spectrum-based error corrector for Illumina sequence data. Bioinformatics 29 (3):308-315. doi:10.1093/bioinformatics/bts690

61. Salmela L (2010) Correction of sequencing errors in a mixed set of reads. Bioinformatics 26 (10):1284-1290. doi:10.1093/bioinformatics/btq151

62. Koren S, Schatz MC, Walenz BP, Martin J, Howard JT et al. (2012) Hybrid error correction and de novo assembly of single-molecule sequencing reads. Nat Biotechnol 30 (7):693-700. doi:10.1038/nbt.2280

Chapter 10
Assessment of Next-Generation Sequence Assembly

Abstract Although there are different measures to evaluate assembler performance and assembly quality, developing assessment tools that incorporate present measures and defining new ones for the various assembly types (genomic, transcriptomic, and metagenomic) still remain a major challenge in the next-generation environment. In this chapter, we will introduce different approaches for assembly assessment as well as discuss upcoming assembly evaluation studies/tools.

10.1 Introduction to Assembly Assessment

The assessment of the assembly process is mainly performed from two perspectives. The first perspective is assembly quality, which evaluates the contiguity, consistency, and accuracy of the assembled genomes using different approaches [1–4]. The second perspective is the performance and usability of the assembler, which includes numerous issues such as hardware and software requirements, ease of installation and execution, user-friendly interfaces, run time per analysis, required memory per 1 GB of data, and the speed of responsiveness to user commands [5–8].

10.2 Contiguity and Consistency Measures

10.2.1 Contiguity Assessment

Statistics metrics are usually used to assess the contiguity of the assembled contigs/scaffolds. These metrics include the distribution of their lengths, their maximum, minimum and average lengths, the number of resulting contigs/scaffolds, the total sum of the assembled contigs/scaffolds, the total length of their short reads, and the N_x score. N_{50} and N_{75} represent the most important metrics for measuring contig/

S. El-Metwally et al., *Next Generation Sequencing Technologies*
and Challenges in Sequence Assembly, SpringerBriefs in Systems Biology 7,
DOI 10.1007/978-1-4939-0715-1_10, © The Authors 2014

scaffold contiguity. They are defined as the length of the contig/scaffold such that 50 %/75 % of its bases are in contigs of greater or equal length [1–4, 9–12]. Although a large value of the N_x score indicates more contiguity in the assembled contigs/scaffolds, the misassembly of contig/scaffold sequences may also increase the score [13].

10.2.2   Consistency Assessment

Due to the presence of abundant information in paired-end libraries, including the estimation of insert size among each pair of reads and their orientation, approaches assessing consistency can utilize this information in the evaluation process. Following the completion of the assembly process, read pairs can be located in the draft sequence. In this case, a comparison of the assembly process with the annotated information of the read pairs (such as separation distance or orientation) can occur. Based on the number of satisfying constraints, we can infer the validity of the assembled sequence [14]. A recently introduced metric also utilizes the idea of aligning the paired-end reads to the assembled genome in generating Feature-Response Curves (FRC) to overcome the available tradeoff between the contiguity and accuracy of the assembly results [15, 16]. Other consistency methods target the type of sequence being assembled (such as haplotype sequences) [3] as well as the constraints imposed by the read coverage to assess the assembled sequences [17] or optical maps [18].

10.3   Accuracy Measures

Comparing the draft sequence assemblies to ones that have been completed represents the most important metric in evaluating the assembly quality [3, 9]. This reference can be an assembled genome of the same species or of a closely related species. The comparative process takes different perspectives such as aligning the two sequences using one of the available alignment tools (i.e., tools that were mentioned in Chap. 2) that report the percentage covered by the assembled sequence [5, 19], the long-range contiguity of the assembled contigs/scaffolds [20], their accuracy and the introduction of modification patterns in the assembled sequences such as insertions, deletions, and substitutions [21]. Furthermore, the comparison process assists in the identification of core genetic components and novel genes [22]. The number of misassembled contigs/scaffolds (i.e., breaks) and the number of mis-aligned bases (i.e., mis-calls) are also used as accuracy metrics in the context of alignment to a reference sequence [23]. Another perspective for assessment occurs during the unavailability of the reference genome. In this case, the comparative process requires independent genetic material from a public database. These genetic components (such as mRNA or cloned genes) can only be utilized if they and the assembled sequences belong to the same type of organism. When this criterion cannot be met, the accuracy approaches enlist components from closely related organisms or conserved sequences [1, 22].

10.4 Assembler's Performance Measures

The runtime and memory usage of an assembler are the most important criteria for the usability measure. Depending on the available computational resources, current assemblers used in next-generation environments are classified into two categories. In the first category, the assemblers run on a single machine with very large memory requirements, e.g., to assemble human and mammalian genomes [19, 24]. In the other category, assemblers are run on tightly coupled cluster machines [25]. The high-throughput nature of next-generation sequencing technologies and the presence of short-read sequences and their quality scores imposes a major constraint on the system memory available. To ensure efficient memory savings, most assemblers formulate the assembly problem as a set of graph nodes and rely on efficient data structures to accommodate these nodes. These different graph models were discussed earlier (see Sect. 9.3), including their advantages and disadvantages with respect to computational resources and several studies that reformulated their representations to ensure efficient storage in memory. However, no memory-efficient solution is presently available for next-generation sequence assemblers, creating a need for new tools and algorithms in this area.

10.5 Assessment Tools and Evaluation Studies for Assessing Assembly Quality

There are several studies for evaluating assembly quality based on combining the approaches that we have discussed previously or defining novel strategies. Furthermore, there are tools that are especially designed for the assessment of the sequence assembly quality. However, the generation of assessment tools that consider the complexity of the data sets being assembled, the assembly algorithms, different parameter settings, and the nature of sequencing experiments are still lacking [21, 26]. It is also important to note that there is always a tradeoff between the different quality measures. For instance, trying to maximize the value of one measure (i.e., improve contigs/scaffolds connectivity) may decrease the value of another (i.e., contigs/scaffolds accuracy). Here, we will mention some studies that attempted to design assessment approaches and metrics that are applicable to wide range of next-generation sequence assembly techniques. Then, we will review the available assembly assessment tools.

10.5.1 Evaluation Studies for Assessing Assembly Quality

Assemblathon [27] is one of the studies that defined its own statistical metrics in addition to existing ones. It uses the haplotype sequences as reference measures to newly defined metrics such as NG_{50}, which is the same as N_{50} but uses an average length of haplotype sequences instead of contig/scaffold lengths during its computation. Similarly, $CPNG_{50}/SPNG_{50}$ denotes the average length of contigs/scaffolds consistent with the haplotype sequences, while CC50 measures the connectivity between any two randomly chosen points in the assembled genomes. The recently published version of the Assemblathon [28] addressed some practical issues during assembly evaluation, including the consideration of diverse assembly results from various assemblers with different parameter settings, the choice of assemblers based on metrics of interest and overlooking contiguity metrics when studying the genetic components of the assembled sequences.

E-size is yet another statistical metric introduced in GAGE [13]. E-size measures the expectation that a certain point (or base), which is chosen randomly from a reference genome, is located in the assembled contigs/scaffolds in terms of their lengths. Additionally, GAGE also discussed the different factors that can affect the evaluation process, such as the complexity of the genome being assembled and the employed assembler. It also reported that various statistical measures cannot be used alone in indicating assembly quality due to inefficiencies in representing the contiguity and accuracy of the assembled sequences. A more recent version of this study is called GAGE-B [29]. GAGE-B evaluated different bacterial genome assemblers using libraries with high coverage reads and studied the effect of the coverage and read lengths on the assembly quality.

Additionally, Haiminen et al. [30] reported that the assessment process can be affected by the nature of sequencing experiments, such as the average length of short reads, their coverage, and the rate of sequencing errors. Furthermore, they give a different score for each mis-call base according to diverse-modified operations, such as substitutions, insertions, deletions, reordering, redundancy, and relocations. The accuracy of the assembled sequence is determined by gathering these scoring values.

10.5.2 Assembly Assessment Tools

QUAST [31] is an assessment tool that uses a combination of metrics which consider the presence or absence of the reference genomes. It uses N_{50}, NG_{50}, NA_{50}, and NGA_{50} in measuring the assembly quality in terms of aligned blocks rather than aligned contigs/scaffolds. QUAST also combines other discussed metrics such as the total number of misassembled contigs/scaffolds and genetic components. Moreover, it provides a full set of functionality to generate different statistical reports supplemented with plots and figures.

Computing Genome Assembly Likelihoods (CGAL) [32] introduced the likelihood metric during de novo assembly evaluation based on the uniformity of the read coverage, errors in the sequenced reads, the distribution of insert sizes, and the size of unassembled reads.

REAPR [33] is another reference-free assessment tool that identifies errors in the assembled sequences using paired-end reads and provides useful information to the end users that reflects the quality of the algorithm used in the assembly process.

10.6 Assessment of Transcriptome and Metagenomes Assembly Quality

The assessment of assembled transcripts also represents a challenge in the next-generation environment since it relies on the abundance of reference transcripts, its length, its different splicing isoforms, and the existence of novel transcripts. Martin and Wang proposed different metrics for assessing transcriptome assembly at different levels of complexity in the context of the abundance of reference transcripts that are well expressed and originate from the same transcriptome sequences [34]. These metrics include accuracy, completeness, contiguity, chimerism, and variant resolution. Although these metrics measure the assembled transcripts according to a set of reference transcripts, they provide useful insight regarding the correct number of assembled bases, the percentage of coverage with respect to reference transcripts, the number of chimeric transcripts that are introduced during the assembly process, and the percentage of resulting variations in the assembled transcripts [34, 35]. If the reference transcripts are not available, other complementary approaches may be utilized instead. This includes examining the encoding of full-length ORFs in different isoforms and performing subsequent validation through the use of proteomic assays [36].

The evaluation of metagenomic sequence assemblies is another formidable challenge in the next-generation sequencing environment due to the presence of a variety of genetic materials from different microbial communities. Mende and colleagues [37] proposed a number of metrics for evaluative purposes, including the number of chimeric contigs, the accuracy of contigs based on their defined scoring scheme, and the variety of genetic components in the resulting assembly sequences.

Charuvake and Rangwala [38] presented the entropy metric to measure the degree of chimerism in contig sequences. Furthermore, they exploited the paired-end reads and sequence coverage to measure the assembly quality. Recently, Assembly Likelihood Evaluation (ALE) [39] announced a reference independent framework for assessing metagenomic and single-cell assemblies. ALE utilizes statistical methods that rely on different informational sources such as paired-end constraints and relevant factors during sequencing experiments (i.e., coverage, errors, and length). In addition, it reports various assembly errors such as base-call errors, misassembled chimeric sequences, as well as genome rearrangements that are a result of indel operations.

References

1. Church DM, Goodstadt L, Hillier LW, Zody MC, Goldstein S et al. (2009) Lineage-specific biology revealed by a finished genome assembly of the mouse. PLoS Biol 7 (5):e1000112. doi:10.1371/journal.pbio.1000112
2. Colbourne JK, Pfrender ME, Gilbert D, Thomas WK, Tucker A et al. (2011) The ecorespon-sive genome of Daphnia pulex. Science 331 (6017):555-561. doi:10.1126/science.1197761
3. Lindblad-Toh K, Wade CM, Mikkelsen TS, Karlsson EK, Jaffe DB et al. (2005) Genome sequence, comparative analysis and haplotype structure of the domestic dog. Nature 438 (7069):803-819. doi:nature04338

4. Locke DP, Hillier LW, Warren WC, Worley KC, Nazareth LV et al. (2011) Comparative and demographic analysis of orang-utan genomes. Nature 469 (7331):529-533. doi:10.1038/nature09687

5. Zhang W, Chen J, Yang Y, Tang Y, Shang J et al. (2011) A practical comparison of de novo genome assembly software tools for next-generation sequencing technologies. PLoS One 6 (3):e17915. doi:10.1371/journal.pone.0017915

6. Alkan C, Sajjadian S, Eichler EE (2011) Limitations of next-generation genome sequence assembly. Nat Methods 8 (1):61-65. doi:10.1038/nmeth.1527

7. Golovko G, Khanipov K, Rojas M, Martinez-Alcantara A, Howard JJ et al. (2012) Slim-Filter: an interactive windows-based application for illumina genome analyzer data assessment and manipulation. BMC bioinformatics 13:166. doi:10.1186/1471-2105-13-166

8. Powell DR, Seemann T (2013) VAGUE: a graphical user interface for the Velvet assembler. Bioinformatics 29 (2):264-265. doi:10.1093/bioinformatics/bts664

9. Li R, Fan W, Tian G, Zhu H, He L et al. (2010) The sequence and de novo assembly of the giant panda genome. Nature 463 (7279):311-317. doi:10.1038/nature08696

10. Lin Y, Li J, Shen H, Zhang L, Papasian CJ et al. (2011) Comparative studies of de novo assembly tools for next-generation sequencing technologies. Bioinformatics 27 (15):2031-2037. doi:10.1093/bioinformatics/btr319

11. Liu Y, Qin X, Song XZ, Jiang H, Shen Y et al. (2009) Bos taurus genome assembly. BMC genomics 10:180. doi:10.1186/1471-2164-10-180

12. Ming R, Hou S, Feng Y, Yu Q, Dionne-Laporte A et al. (2008) The draft genome of the transgenic tropical fruit tree papaya (Carica papaya Linnaeus). Nature 452 (7190):991-996. doi:10.1038/nature06856

13. Salzberg SL, Phillippy AM, Zimin A, Puiu D, Magoc T et al. (2012) GAGE: A critical evaluation of genome assemblies and assembly algorithms. Genome research 22 (3):557-567. doi:10.1101/gr.131383.111

14. Huson DH, Halpern AL, Lai Z, Myers EW, Reinert K et al. Comparing Assemblies Using Fragments and Mate-Pairs. In: WABI '01 Proceedings of the First International Workshop on Algorithms in Bioinformatics Århus, Denmark, 2001. Springer Berlin Heidelberg, pp 294-306

15. Narzisi G, Mishra B (2011) Comparing de novo genome assembly: the long and short of it. PLoS One 6 (4):e19175. doi:10.1371/journal.pone.0019175

16. Vezzi F, Narzisi G, Mishra B (2012) Feature-by-feature—evaluating de novo sequence assembly. PLoS One 7 (2):e31002. doi:10.1371/journal.pone.0031002

17. Phillippy AM, Schatz MC, Pop M (2008) Genome assembly forensics: finding the elusive misassembly. Genome Biol 9 (3):R55. doi:10.1186/gb-2008-9-3-r55

18. Zhou S, Bechner MC, Place M, Churas CP, Pape L et al. (2007) Validation of rice genome sequence by optical mapping. BMC genomics 8:278. doi:1471-2164-8-278

19. Li R, Zhu H, Ruan J, Qian W, Fang X et al. (2010) De novo assembly of human genomes with massively parallel short read sequencing. Genome research 20 (2):265-272. doi:10.1101/gr.097261.109

20. Gnerre S, Maccallum I, Przybylski D, Ribeiro FJ, Burton JN et al. (2011) High-quality draft assemblies of mammalian genomes from massively parallel sequence data. Proceedings of the National Academy of Sciences of the United States of America 108 (4):1513-1518. doi:10.1073/pnas.1017351108

21. Meader S, Hillier LW, Locke D, Ponting CP, Lunter G (2010) Genome assembly quality: assessment and improvement using the neutral indel model. Genome research 20 (5):675-684. doi:10.1101/gr.096966.109

22. Parra G, Bradnam K, Ning Z, Keane T, Korf I (2009) Assessing the gene space in draft genomes. Nucleic acids research 37 (1):289-297. doi:10.1093/nar/gkn916

23. Hubisz MJ, Lin MF, Kellis M, Siepel A (2011) Error and error mitigation in low-coverage genome assemblies. PLoS One 6 (2):e17034. doi:10.1371/journal.pone.0017034

24. Li H (2012) Exploring single-sample SNP and INDEL calling with whole-genome de novo assembly. Bioinformatics 28 (14):1838-1844. doi:10.1093/bioinformatics/bts280

25. Simpson JT, Wong K, Jackman SD, Schein JE, Jones SJ et al. (2009) ABySS: a parallel assembler for short read sequence data. Genome research 19 (6):1117-1123. doi:10.1101/gr.089532.108

26. Salzberg SL, Yorke JA (2005) Beware of mis-assembled genomes. Bioinformatics 21 (24):4320-4321. doi: 10.1093/bioinformatics/bti769

27. Earl D, Bradnam K, St John J, Darling A, Lin D et al. (2011) Assemblathon 1: a competitive assessment of de novo short read assembly methods. Genome research 21 (12):2224-2241. doi:10.1101/gr.126599.111

28. Bradnam KR, Fass JN, Alexandrov A, Baranay P, Bechner M et al. (2013) Assemblathon 2: evaluating de novo methods of genome assembly in three vertebrate species. Gigascience 2 (1):10. doi:2047-217X-2-10

29. Magoc T, Pabinger S, Canzar S, Liu XY, Su Q et al. (2013) GAGE-B: an evaluation of genome assemblers for bacterial organisms. Bioinformatics 29 (14):1718-1725. doi:10.1093/bioinformatics/btt273

30. Haiminen N, Kuhn DN, Parida L, Rigoutsos I (2011) Evaluation of methods for de novo genome assembly from high-throughput sequencing reads reveals dependencies that affect the quality of the results. PLoS One 6 (9):e24182. doi:10.1371/journal.pone.0024182

31. Gurevich A, Saveliev V, Vyahhi N, Tesler G (2013) QUAST: quality assessment tool for genome assemblies. Bioinformatics 29 (8):1072-1075. doi:10.1093/bioinformatics/btt086

32. Rahman A, Pachter L (2013) CGAL: computing genome assembly likelihoods. Genome Biol 14 (1). doi: 10.1186/Gb-2013-14-1-R8

33. Hunt M, Kikuchi T, Sanders M, Newbold C, Berriman M et al. (2013) REAPR: a universal tool for genome assembly evaluation. Genome Biol 14 (5):R47. doi:gb-2013-14-5-r47

34. Martin JA, Wang Z (2011) Next-generation transcriptome assembly. Nature reviews Genetics 12 (10):671-682. doi:10.1038/nrg3068

35. Martin J, Bruno VM, Fang Z, Meng X, Blow M et al. (2010) Rnnotator: an automated de novo transcriptome assembly pipeline from stranded RNA-Seq reads. BMC genomics 11:663. doi:10.1186/1471-2164-11-663

36. Adamidi C, Wang Y, Gruen D, Mastrobuoni G, You X et al. (2011) De novo assembly and validation of planaria transcriptome by massive parallel sequencing and shotgun proteomics. Genome research 21 (7):1193-1200. doi:10.1101/gr.113779.110

37. Mende DR, Waller AS, Sunagawa S, Jarvelin AI, Chan MM et al. (2012) Assessment of metagenomic assembly using simulated next generation sequencing data. PLoS One 7 (2):e31386. doi:10.1371/journal.pone.0031386

38. Charuvaka A, Rangwala H (2011) Evaluation of short read metagenomic assembly. BMC genomics 12 Suppl 2:S8. doi:10.1186/1471-2164-12-S2-S8

39. Clark SC, Egan R, Frazier PI, Wang Z (2013) ALE: a generic assembly likelihood evaluation framework for assessing the accuracy of genome and metagenome assemblies. Bioinformatics 29 (4):435-443. doi:10.1093/bioinformatics/bts723

Chapter 11
Next-Generation Sequence Assemblers

Abstract There are several assemblers employed in next-generation environment. These may be classified according to their respective graph construction approaches (see Chap. 9) or their targeted data sets. In this chapter, we will present select examples of next-generation sequence assemblers and discuss their implementation approaches. The assemblers we discuss have been selected carefully to represent the available assembly approaches and the first four stages of the assembly process. Tools related to assembly assessment, the fifth stage, have been previously discussed in Chap. 10.

11.1   Introduction

Assemblers in the next-generation environment can be classified according to their targeted data sets. Since the targeted next-generation data can be either genomic or transcriptomic; assemblers of these types of information can be classified into genome assemblers and transcriptome assemblers respectively [1]. Furthermore, metagenomic assemblers that are specifically designed to assemble next-generation metagenomic data have also been developed, but are not discussed here due to text limitations.

From an implementation point of view, assemblers can be classified according to their approaches to the graph construction process (see Chap. 9). In this respect, we can have overlap-based assemblers, k-spectrum-based assemblers, greedy-based assemblers, and hybrid assemblers [1]. Furthermore, the distinct technical properties of each assembler can be used to define a technical classification based on the assembler's platform, licencing options (free or commercial), availability of source code, and other such criteria.

Here, we will follow the implementation classification of the assemblers and attempt to discuss most of the available implementation approaches. We will mainly focus on graph construction approaches in relation to each assembler and the distinct methods employed in transforming the graph representations into contigs. Moreover, we present two tables that list the practical (Table 11.1) and technical

S. El-Metwally et al., *Next Generation Sequencing Technologies*
and Challenges in Sequence Assembly, SpringerBriefs in Systems Biology 7,
DOI 10.1007/978-1-4939-0715-1_11, © The Authors 2014

Table 11.1 Practical features in next-generation sequence assemblers

Assemblers	Sequencing platform	Input file format	Output file format	Genome/ transcriptome	Prokaryotic/eukaryotic	Single/ paired-end reads
Newbler[a]	Any platform	.sff, .fasta, .qual	.fna, .qual, .txt, .sff, .tsv, .ace	Genome[T]	Prokaryotic	S/P
GS de novo assembler						
Edena	Illumina/Solexa	.fasta[b]	.fasta[b], .cov, .info	Genome	Prokaryotic	S/P
Celera[a]	Sanger, Illumina/Solexa, 454, Ion Torrent, Pacific Biosciences	.fasta	.asm	Genome	Eukaryotic/prokaryotic	S/P
CABOG		.fastq	.fasta .posmap, .qc			
wgs-assembler		.frg, .sff				
Shorty	SOLiD	.fasta	.fasta	Genome	Prokaryotic	S/P
Forge[a]	Illumina/Solexa, Helicos	.fasta, .fastq, .qual	.fasta, .txt	Genome	Eukaryotic/prokaryotic	S/P
SGA[a]	Hybrid of Sanger, 454 and Illumina/Solexa	.fastq	.fasta	Genome	Eukaryotic/prokaryotic	S/P
Readjoiner[a]	Illumina/Solexa[c]	.fasta, .fastq	.fasta, .dot, .sga	Genome	Eukaryotic/prokaryotic	S/P
Fermi	Illumina/Solexa	.fastq	.fastq-like format	Genome	Eukaryotic/prokaryotic	S/P
Euler-SR	454, Illumina/Solexa	.sff[b], .fastq, .eland	.fasta[b]	Genome	Eukaryotic/prokaryotic	S/P
ALLPATHS-LGa	Illumina/Solexa, Pacific Biosciences	.fastb, .qualb, .pairs	.fasta, .efasta	Genome	Prokaryotic/eukaryotic	S/P
Velvet[a]	454, Illumina/Solexa, SOLiD	.fasta .fastq .fasta.gz fastq.gz .sam, .bam, .eland, .gerald	.fasta, .afg, .txt	Genome	Prokaryotic/eukaryotic	S/P
ABySS	Illumina/Solexa, 454, SOLiD	.fastqm, .fasta, .qseq, .export, .sam, .bam	.fasta, .hist, .dot, .adj .dist, .path, coverage.hist	Genome[T]	Eukaryotic/prokaryotic	S/P
SOAPdenovo	Illumina/Solexa	.fastq, .fasta	.contig .scafSeq	Genome[T]	Eukaryotic/prokaryotic	S/P

Tool	Sequencing platforms	Input formats	Output formats	Type	Organism	S/P
SparseAssembler[a]	Illumina/Solexa	.fasta, .fastq	.fasta	Genome	Eukaryotic/prokaryotic	S
SSAKE	Illumina/Solexa	.fasta, raw	.fasta [S]	Genome	Eukaryotic/prokaryotic	S/P
SHARCGS[a]	Illumina/Solexa	.fasta, raw	.fasta	Genome	Eukaryotic/prokaryotic	S
Vcake[a]	Illumina/Solexa	.fasta, raw	.fasta	Genome	Eukaryotic/prokaryotic	S
QSRA	Illumina/Solexa	.fasta, .raw	.fasta[S]	Genome	Prokaryotic	S
Taipan[a]	Illumina/Solexa	.raw	.fasta	Genome	Eukaryotic/prokaryotic	S
Rnnotator[a]	Illumina/Solexa	.fastq	.fastq	Transcriptome	Prokaryotic	S
Oases[a]	454, Illumina/Solexa, SOLiD	fasta, .fastq, sam/bam	.fasta.txt	Transcriptome	–	S/P
Trinity[a]	Illumina/Solexa	.fasta, .fastq	.fasta	Transcriptome	–	S/P
Ray[a]	454, Illumina/Solexa	fasta, .fa .fasta.gz .fa.gz. fasta.bz2 fa.bz2 .fq.gz fastq.fq fastq.gz .fq.gz fastq .bz2 fq.bz2 export.txt qseq.txt .sff	.fasta .txt AMOS.afg	Genome[M]	Eukaryotic/prokaryotic	S/P
IDBA[a]	Illumina/Solexa	.fasta .fastq	.fasta	Genome[T][M][C]	Eukaryotic/prokaryotic	S/P
MIRA	Sanger, Illumina/Solexa, 454, Ion Torrent, Pacific Biosciences	.exp .caf .fasta .fastq .phd.scf	.fasta .html .caf .gap4 .acc .tcs	Genome[T]	Eukaryotic/prokaryotic	S/P
MaSurCa	Sanger, Illumina/Solexa, 454	.fastq .frg	.fasta	Genome	Eukaryotic/prokaryotic	S/P
SPAdes[a]	Sanger, Illumina/Solexa, Ion Torrent, Pacific Biosciences	.fasta .fastq	.fasta .fastg	Genome[C][D]	Eukaryotic/prokaryotic	S/P
Minia[a]	Illumina/Solexa	.fasta	.fasta	Genome	Eukaryotic/prokaryotic	S

[a]Personal communications with authors
[b]Users experiences and communities websites
[c]Available for other sequencing platforms if the datasets are filtered
[T]Transcriptome assembly version is available
[S]Speculated, based on sequencing platforms
[M]Metagenome assembly version is available
[C]Single cell assembly version is available
[D]Diploid genome assembly version is available

(Table 11.2) features of about 30 different assemblers. Further information related to assembler implementation, classification, and a comparison of features may be found in El-Metwally et al. [1].

11.2   Next-Generation Genome Assemblers

11.2.1   Overlap-Based Assemblers

Newbler [2] is the assembler distributed by 454 Life Sciences and follows the OLC graph principle. It assembles short reads during two phases of the graphing process. The first phase generates mini contigs (unitigs) by combining overlapping reads. These unitigs are used as bases to generate larger contigs in the second phase of graph construction. This phase uses Multialigner, which creates a layout contig by implementing pairwise alignments between the unitigs, and subsequently, produces a consensus sequence. Furthermore, Newbler creates a layout and consensus in a "flow space" using platform-based call signals associated with each nucleotide. The recent versions differ in their names and descriptions in comparison to the published algorithm.

Edena [3] is an OLC graph assembler for reads of uniform length from Illumina. The first key step by Edena involves the preprocessing of the redundancy of the reads without loss of information. Owing to oversampling of the reads, Edena retains a single copy of each read. In this case, reads that contain (ambiguous) unresolved bases are discarded due to the fact that Edena operates with exact matching overlaps between the reads. Furthermore, the frequency of each read is recorded for computing coverage depth in the contigs for quality purposes. In the second step, Edena computes the overlap by indexing the reads in a suffix array. The set of computed overlaps are structured in a bidirectional graph, where each read represents a node and two overlapped reads (nodes) are connected by a bidirectional edge. The nodes in the graph are traversed in two directions corresponding to the reads and their reverse complements to build a valid assembly. The third step is the graph simplification process, which reduces the graph complexity and repairs its unresolved bubbles and spurs. After the graph simplification operations are completed, the contigs are produced by concatenating the simple paths on the graph.

The Celera [4] assembler implements the same methodology adopted by BLAST [5]. This involves the seed-and-extend algorithm and consists of several phases: (1) a screener phase which filters reads that are repeats from other reads, (2) an overlapper phase which compares each read with all other reads to detect suffix-to-prefix overlaps, (3) a unitiger phase that combines reads by their overlaps to form mini-assemblies called unitigs, and (4) a scaffolder phase that combines unitigs to form scaffolds (supercontigs). In the final step, Celera computes the optimal consensus sequence by aligning reads according to consensus metrics and consensus base-calls. The Celera assembler was originally developed for Sanger reads but recent releases have also included the ability to assemble reads from the 454, Illumina, Pacific Biosciences, and Ion Torrent platforms.

Table 11.2 Technical features in next-generation sequence assemblers

Assemblers	Operating system	Programming language	Single PC/cluster	Open source	Website	Reference
Newbler[a] (V2.8) GS de novo assembler	Linux (32–64) bits CentOS or RedHat	C++	Single	N	http://454.com/contact-us/software-request.asp	[2]
Edena (V3.121122)	Linux (32–64) bits Windows	C++	Single	Y	http://www.genomic.ch/edena	[3]
Celera[a] (V7.0) CABOG wgs-assembler	Linux/Unix (64 bits) Mac OS X, Darwin, FreeBSD	C++/C/Perl	Single/cluster	Y	http://wgs-assembler.sourceforge.net/	[4, 6]
Shorty (V2.0)	Windows, Linux, Mac OS X[b]	C++	Single	Y	http://www.cs.sunysb.edu/~skiena/shorty	[7]
Forge[a]	Windows, Linux, Mac	C++	Single/cluster	Y	http://combiol.org/forge/	[8]
SGA[a]	Linux, Mac OS X	C++	Single/cluster	Y	https://github.com/jts/sga	[9]
Readjoiner[a] (V1.2)	Linux (32–64) bits, Mac OS X, Cygwin, POSIX-compatible	C	Single/cluster	Y	http://www.zbh.uni-hamburg.de/readjoiner	[11]
Fermi	Linux	C	Single	Y	https://github.com/lh3/fermi	[39]
Euler-SR	Linux (32–64) bits[b]	C++/Perl[b]	Single[b]	N/A[b]	N/A	[14]
ALLPATHS-LG[a]	Linux (64 bits)	C++	Single	Y	http://www.broadinstitute.org/software/allpaths-lg/blog	[18–20]
Velvet[a] (V1.2.08)	Linux (32–64) bits, Mac OS X, Cygwin, Sparc/Solaris	C	Single	Y	http://www.ebi.ac.uk/~zerbino/velvet	[21]

(continued)

Table 11.2 (continued)

Assemblers	Operating system	Programming language	Single PC/cluster	Open source	Website	Reference
ABySS (V1.3.4)	For all platforms	C++	Single/Cluster	Y	http://www.bcgsc.ca/platform/bioinfo/software/abyss	[23]
SOAPdenovo (V1.05)	Linux (32–64), Mac[b]	C/C++[b]	Single	Y	http://soap.genomics.org.cn/soapdenovo.html	[24]
SparseAssembler[a]	Linux (64) bits	C/C++	Single	Y	http://sites.google.com/site/sparseassembler/	[27]
SSAKE (V3.8)	For all platforms	Perl	Single	Y	http://www.bcgsc.ca/bioinfo/software/ssake	[40]
SHARCGS[a]	Linux/Unix (32–64) bits	Perl	Single	Y	http://sharcgs.molgen.mpg.de/	[41]
Vcake[a]	Windows (32–64) bits; Linux/Unix (32–64) bits	Perl/C	Single	Y	http://sourceforge.net/projects/vcake/	[30]
QSRA (V1.0)	Linux/Unix (32–64) bits	C++	Single	Y	http://mocklerlab.org/tools/2	[31]
Taipan[a]	Linux	C	Single	Y	http://taipan.sourceforge.net	[32]
Rnnotator[a] (V 3.3.0)	Linux (64) bits	Perl	Single/cluster	N	https://sites.google.com/a/lbl.gov/rnnotator/home	[35]
Oases[a] (V0.2.08)	Linux (32–64) bits[c]	C/Python	Single	Y	http://www.ebi.ac.uk/~zerbino/oases/	[37]
Trinity[b]	Linux	C++/java/Perl	Single/cluster	Y	http://trinityrnaseq.sourceforge.net/	[38]
Ray[a]	POSIX	C++	Single/cluster	Y	http://denovoassembler.sourceforge.net/	[42]
IDBA[a]	Linux (64) bits	C++	Single/cluster	Y	http://i.cs.hku.hk/~alse/hkubrg/projects/idba/	[43–45]
MIRA	Linux, Mac OS X	C/C++	Single	Y	http://sourceforge.net/projects/mira-assembler/	[46, 47]
MaSurCa	Linux (64) bits	C++/Perl	Single/cluster	Y	ftp://ftp.genome.umd.edu/pub/MaSuRCA/	[48]
SPAdes[a]	Linux (64) bits, Mac OS X	C++/Python	Single/cluster	Y	http://bioinf.spbau.ru/en/spades	[49]
Minia	Linux	C++	Single	Y	http://minia.genouest.org/	[50]

[a]Personal communications with authors
[b]Users experiences and communities websites
[c]Any system with GCC

The CABOG [6] assembler is a specific variant of the Celera assembler [4] for 454 short reads. It begins the assembly process through a filtering step which removes the reads that contain at least one unknown base, N, and trims the reads that have partial overlaps which do not span the ends of the reads being assembled. Subsequently, the anchor and overlap step utilizes a seed-and-extending algorithm to find the exact match (overlapping) between the compressed reads. Each seed represents a k-mer and the number of occurrences of a unique k-mer is recorded to differentiate repetitive k-mers from the others. Each read is treated as a reference and detects all other reads that share a suffix-to-prefix overlap with it by calculating pairwise alignment among them. The reads and overlaps are represented in a graph in which each read is indicated by a pair of nodes with an undirected edge between them. The pair of nodes represents the two pair ends of the read. The directed edge is the one that links two reads that have a suffix-to-prefix overlap.

Shorty [7] is a paired-end short read assembler for SOLiD that also uses the idea of a seed extending algorithm. Short read data sets are organized into a compressed trie data structure to allow the retrieval of reads easily. The pairs are inserted into the trie based on their left and right k-mers. Each node in the trie represents a k-mer, and Shorty can retrieve pairs of reads that contain this k-mer in their left or right ends by searching the trie for k-mers. Each k-mer is used as a seed and is useful for determining a set of neighbors in one position of the assembly process. The set of reads that share one k-mer is plotted in the graph where each node represents a read and each edge denotes overlaps between the reads. The greedy algorithm is used to traverse nodes to find the longest path in the graph that visits all nodes exactly once. The generated path represents a contig, and only paths with maximum length are kept as solution contigs for the assembly problem. The set of contigs resulting from each group is used again as seeds and for iteratively building a graph, where each node represents a contig and each edge signifies an overlap between the contigs. The graph is greedily traversed again to find a maximal path that visits all nodes exactly once. This process is repeated until no more contigs exist. Scaffolds have also been built using the same idea of traversing a graph for contigs.

Forge [8] is a hybrid approach assembler that assembles reads from the Sanger, Illumina, and 454 platforms. It has adaptive algorithms that are able to change their performance according to the characteristics of the reads being assembled, which may have diverse lengths and error models resulting from the different sequencing technologies. The first step is the filtering and trimming of reads based on their length and the phred quality scores of their bases. Forge utilizes a classical OLC graph to find a solution for the assembly problem and traverses graph nodes to locate contigs and scaffolds.

SGA [9] is a string graph assembler that exploits the FM-index data structure (see Chap. 2) [10] to store and correct sequence reads efficiently. Furthermore, the FM-index is used to detect transitive edges early, which directly simplifies the construction of string graphs. The graph is cleaned by removing tips, resolving bubbles, and subsequently traversed to produce contigs and scaffolds.

Readjoiner [11] is another string graph assembler with three pipeline steps: (1) prefilter, which involves the removal of reads that contain ambiguous bases and partial overlaps, and the correction of reads that contain sequencing errors, (2) over-lapper, which comprises of computing all irreducible suffix-to-prefix matches between reads and the indexing of an appropriate subset of them for efficient storage and retrieval, and (3) assembler, which involves the building of a string graph of reads, the cleaning of tips and bubbles, and the traversing of the graph to build contigs.

11.2.2   K-Spectrum-Based Assemblers

Euler-SR [12–16] is a short read De Bruijn graph assembler that implements a pre-processing filter to filter erroneous reads based on their k-mers. The k-mers with high frequencies are trusted ones and become candidates for correcting less-frequent ones. Euler-SR builds De Bruijn graphs of different k-mer sizes, and compares them to detect missing edges that are used to enlarge contigs in the assembly. Then, it applies heuristics to simplify and repair the graphs, and incorporates the A-Bruijn graph [17] as a repeated graph and classifies repeats according to their similarities. Furthermore, Euler-SR uses paired-ends information to resolve long repeats in a process called mate threading, and identifies erroneous links by aligning paired-ends against a De Bruijn graph. The maximum insert size between the two paired-end reads is used to detect the correct/incorrect paths between them. Euler-SR is used to assemble reads from the 454 and Illumina platforms with paired and unpaired read libraries.

ALLPATHS-LG [18–20] uses the same concept as the other De Bruijn graph assemblers but has its own preprocessing filter based on the k-spectrum approach. The frequency of each k-mer is recorded and used as a basis to filter erroneous reads. For each consecutive k-mer in the genome, a corresponding number is assigned to it. Subsequently, the set of k-mer intervals is computed, which corre-sponds to the k-mer paths in the graph. ALLPATHS-LG builds a unipath graph from the k-mer paths and aims to find the maximum branchless intervals of the k-mer numbers. After determining the unipath intervals, these intervals are used to make larger ones by searching their successors and predecessors and joining them together. The process is iterated to produce larger unipaths. The three main func-tions of the ALLPATHS-LG assembler are finding all paths across the graph nodes, choosing suitable ones to form unipaths, and localizing read pairs to detect the iso-lated regions that can assemble them independently. ALLPATHS-LG is an improve-ment over the original ALLPATHS [18, 20] algorithm in assembling large mammalian genomes that have an abundance of repetitive structures. In relation to specific platforms, ALLPATHS-LG is used to assemble reads from the Illumina and Pacific Biosciences systems.

Velvet [21] is a De Bruijn graph assembler that does not implement a preprocess-ing filter, but applies a series of heuristics to simplify and reduce graph complexity. This assembler is a collection of algorithms that aim to correct graph errors and solve graph repeats. It utilizes the basic idea of De Bruijn graph construction and manipulation. Velvet iteratively removes the tips only if they are shorter than 2k and have a lower coverage than other alternative paths on the graph. Furthermore, it can exploit long reads from other sequencing platforms to confirm or reject the paths of low-coverage nodes. Velvet implements the Breadcrumb algorithm that uses paired-end reads to correctly extend contigs in the region of repeats, and integrates the Pebble and Rock Band algorithms to resolve more complex repeated structures [22]. Moreover, Velvet is the only assembler that can handle color space data from the SOLiD platform and can assemble reads from the Illumina and 454 platforms.

ABySS [23] is a multi-threaded assembler that allows a distribution of the De Bruijn graph and a parallel implementation of assembly algorithms across a network of computers. The assembler has two phases: the first phase includes extracting k-spectrum contained in the set of reads, constructing a De Bruijn graph that represents k-mers and their correlation with each other, and deriving initial contigs using path traversing algorithms. The second phase includes building larger contigs (scaffolds) using mate-pairs information and closing the gaps between them. ABySS is used to assemble reads from the Illumina platform.

SOAPdenovo [24] is a short read assembler that is based on the idea of the De Bruijn graph and is used to assemble reads from the Illumina platform. This assembler is integrated into SOAP (short oligonucleotide alignment program) [25] and the package is named SOAPdenovo, which provides useful analysis and detection of genetic variations such as SNPs, small insertions, and deletions in the genome. SOAPdenovo begins by preprocessing reads to detect and correct erroneous ones. Dynamic programming is used to find a candidate solution for correction based on minimum number of changes (hamming distance). To overcome long computational times due to handling large data sets, thread parallelization is used to distribute a set of reads by sharing the same hash table data structure. This process subsequently builds a De Bruijn graph using k-mers contained in the reads. After that, SOAPdenovo simplifies and repairs the graph by removing short tips. Furthermore, SOAPdenovo also removes bubbles, chimeric connections, and resolves tiny repeats. Then, the assembler breaks paths on the graph at repeated regions to produce assembled contigs. These contigs are realigned later and used with paired-ends information to build scaffolds. A recently improved version of the assembler is called SOAPdenovo2, and provides efficient memory usage, the resolving of repeats, and an increase in the assembly quality of large genomic sequences [26].

To overcome the extensive hardware resources needed to manage a De Bruijn graph, a sparse k-mer graph was introduced and implemented in the SparseAssembler [27]. This assembler constructs a sparse graph, simplifies it by removing bubbles and tips, and traverses it to build contigs.

11.2.3   Greedy-Based Assemblers

SSAKE [28] is a prefix tree-based assembler for Illumina reads that greedily assembles next-generation short reads. The first step of the SSAKE assembler is to create a hash table that stores short reads, where the keys are unique reads and values represent the frequency of the reads. Subsequently, the SSAKE constructs a prefix tree to organize reads and their reverse complements. The reads are sorted in the tree according to their frequencies. After that, SSAKE creates a set of possible $3'$ k-mers and uses them as keywords to find possible overlaps between unassembled reads and the current contig extending in $5'$-direction. The read with the maximum overlapping length is joined to the current contig and removed from the hash table and prefix tree. When all the possibilities have been exhausted for the $5'$ contig extension, the complementary strand is used to extend the contig on the $3'$-end. The process is repeated until no more reads are used for the extension.

SHARCGS [29] employs three steps for assembling short reads from Illumina, which are filter, assembler, and merger. The goal of the filter step is to remove erroneous reads based on their frequencies, existence of overlap partners, and their quality scores. Before starting the assembly algorithm, SHARCGS retains only one copy of each read and generates a reverse complement for it. The assembler step is similar to the SSAKE algorithm. The setting of parameters is a critical task during assembly because stricter settings may lead to the removal of some reads, resulting in shorter contigs. On the other hand, more tolerant settings may lead to the incorporation of erroneous reads during the assembly. The merger step combines contigs that result from three different parameter settings, namely weak, medium, and strong. The goal of the merger is to incorporate contigs of varying lengths at different confirmation settings, which shares overlaps between them and combines them into one contig.

VCAKE [30] improves error handling over the SSAKE platform by exhibiting inexact matching during contig extension. Similar to SSAKE, the sequences are efficiently retrieved from a lookup hash table. However, rather than extending the contigs based on the greedy approach or quality values, VCAKE considers all possible overlaps between the contig being extended (or seed) and the unassembled reads. The extension process occurs at a rate of one base at a time and uses a majority of bases among a set of overlapped reads, allowing imperfect matching with a maximum of one base between the overlapped read and the extended contig.

QSRA [31] expands on the VCAKE [30] notion by combining quality scores in case the overlap length is not sufficient during contig extension. This assembler uses minimum user-defined quality scores to find overlaps between the extended contig and unassembled reads.

11.2.4 Hybrid Assemblers

Taipan [32] is a hybrid assembler that combines the attributes of the OLC and De Bruijn graphs. This assembler chooses the most commonly occurring read in the assembly set <R> as a seed S and tries to extend it carefully in the $3'$-direction, one base at a time, until there are inadequate overlaps or repeats. Following the same concept, it extends in the $5'$-direction as well. The following steps are repeated for extending each contig; the set of overlapped reads are retrieved from a hash table and

Taipan creates a directed overlap graph. The created graph is simplified by the removal of associative edges and the detection and resolving of repeats etc. The set of cleaned nodes are analyzed and used to determine a single base for extending the seed. After finishing the assembly of this contig, all reads that are exactly matched are removed from the read set and hash table. The process of extending the seeds is repeated to reconstruct another contig until no more reads are found in the assembly set.

Wang et al. [33] combined reads from the 454, SOLiD, and Illumina platforms and assembled them separately using a suitable assembler for each read data set. This step was called primary assembly and aimed to evaluate the results from different assemblers and determine suitable parameters for hybrid assembly during the second step. Subsequently, hybrid assembly begins by combining the resulting contigs that represent two or three sequencers. These contigs are filtered together to detect and correct systematic biases or errors produced by each sequencer. Finally, these contigs are aligned together to build scaffolds and close the gaps between them.

Cerdeira et al. [34] propose an alternate approach for hybrid assembly through the combination of contigs produced by different assemblers (i.e., Edena and Velvet) and constructed from different graph paradigms such as OLC and De Bruijn. These contigs are then used as the basis to produce larger and more accurate ones.

11.3 Next-Generation Transcriptome Assemblers

Rnnotator [35] is an automated pipeline for transcriptome assembly that begins with prefiltering a set of reads. This is accomplished by removing redundant and correcting low quality reads based on k-mer coverage. Then, the set of filtered reads are entered into the Velvet [21] assembler to continue the assembly. Rnnotator combines contigs from different Velvet assembly runs with various parameter settings using the Minimus2 assembler from the AMOS packages [36] to produce improved results in relation to the assembled transcripts. In the post-processing filtering stage, Rnnotator uses strand-specific sequencing reads to resolve the overlapping transcripts and determine the direction of the assembled transcripts. Furthermore, it aligns reads back to the contigs to detect single base sequencing errors and generate consensus sequences.

Oases [37] is another transcriptome assembler that uses Velvet to generate a set of contigs corresponding to the assembled transcripts. An adaptive version of the Tour Bus algorithm is used to detect bubbles on the graph, which merges low coverage paths with the higher ones when their sequences are sufficiently similar. In addition, Oases removes local edges that have less coverage contribution to outgoing nodes with respect to other neighboring edges, and filters repeated contigs using the coverage threshold. By using single reads or spanning paired-end reads, the estimation distance among contigs can be detected. Hence, contigs can be combined based on the number of supporting reads. The scaffolds are filtered according to different coverage and statistical factors and grouped into clusters according to similarity in gene related components. These clusters are transitively reduced and the assembled transcripts are extracted from the filtered clustered sets.

Trinity [38] is a modular transcriptome assembler that contains three stages: Inchworm, Chrysalis, and Butterfly. The Inchworm step assembles short reads efficiently using a greedy approach for detecting shared k-mers among the reads. The Chrysalis step clusters contigs according to their related components corresponding to different splicing isoforms, and constructs a De Bruijn graph for each cluster. The Butterfly step analyzes various paths on the De Bruijn graphs using reads and paired-end links as a filter to detect different overlaps among transcripts and their diverse splicing sequences.

References

1. El-Metwally S, Hamza T, Zakaria M, Helmy M (2013) Next-generation sequence assembly: four stages of data processing and computational challenges. PLoS Comput Biol 9 (12):e1003345. doi:10.1371/journal.pcbi.1003345
2. Margulies M, Egholm M, Altman WE, Attiya S, Bader JS et al. (2005) Genome sequencing in microfabricated high-density picolitre reactors. Nature 437 (7057):376-380. doi:nature03959
3. Hernandez D, Francois P, Farinelli L, Osteras M, Schrenzel J (2008) De novo bacterial genome sequencing: Millions of very short reads assembled on a desktop computer. Genome research 18 (5):802-809. doi:10.1101/gr.072033.107
4. Myers EW, Sutton GG, Delcher AL, Dew IM, Fasulo DP et al. (2000) A whole-genome assembly of Drosophila. Science 287 (5461):2196-2204
5. Altschul SF, Gish W, Miller W, Myers EW, Lipman DJ (1990) Basic local alignment search tool. J Mol Biol 215 (3):403-410. doi:10.1016/S0022-2836(05)80360-2
6. Miller JR, Delcher AL, Koren S, Venter E, Walenz BP et al. (2008) Aggressive assembly of pyrosequencing reads with mates. Bioinformatics 24 (24):2818-2824. doi:10.1093/bioinformatics/btn548
7. Hossain M, Azimi N, Skiena S (2009) Crystallizing short-read assemblies around seeds. BMC bioinformatics 10 (Suppl 1):S16. doi:10.1186/1471-2105-10-s1-s16
8. DiGuistini S, Liao NY, Platt D, Robertson G, Seidel M et al. (2009) De novo genome sequence assembly of a filamentous fungus using Sanger, 454 and Illumina sequence data. Genome Biol 10 (9):R94. doi:10.1186/gb-2009-10-9-r94
9. Simpson JT, Durbin R (2012) Efficient de novo assembly of large genomes using compressed data structures. Genome research 22 (3):549-556. doi:10.1101/gr.126953.111
10. Simpson JT, Durbin R (2010) Efficient construction of an assembly string graph using the FM-index. Bioinformatics 26 (12):i367-373. doi:10.1093/bioinformatics/btq217
11. Gonnella G, Kurtz S (2012) Readjoiner: a fast and memory efficient string graph-based sequence assembler. BMC bioinformatics 13:82. doi:10.1186/1471-2105-13-82
12. Chaisson M, Pevzner P, Tang H (2004) Fragment assembly with short reads. Bioinformatics 20 (13):2067-2074. doi:10.1093/bioinformatics/bth205
13. Chaisson MJ, Brinza D, Pevzner PA (2009) De novo fragment assembly with short mate-paired reads: Does the read length matter? Genome research 19 (2):336-346. doi:10.1101/gr.079053.108
14. Chaisson MJ, Pevzner PA (2008) Short read fragment assembly of bacterial genomes. Genome research 18 (2):324-330. doi:10.1101/gr.7088808
15. Pevzner PA, Tang H, Waterman MS (2001) An Eulerian path approach to DNA fragment assembly. Proceedings of the National Academy of Sciences of the United States of America 98 (17):9748-9753. doi:10.1073/pnas.171285098
16. Pevzner PA, Tang H (2001) Fragment assembly with double-barreled data. Bioinformatics 17 Suppl 1:S225-233

17. Pevzner PA, Tang H, Tesler G (2004) De novo repeat classification and fragment assembly. Genome research 14 (9):1786-1796. doi:10.1101/gr.2395204
18. Butler J, MacCallum I, Kleber M, Shlyakhter IA, Belmonte MK et al. (2008) ALLPATHS: de novo assembly of whole-genome shotgun microreads. Genome research 18 (5):810-820. doi:10.1101/gr.7337908
19. Gnerre S, Maccallum I, Przybylski D, Ribeiro FJ, Burton JN et al. (2011) High-quality draft assemblies of mammalian genomes from massively parallel sequence data. Proceedings of the National Academy of Sciences of the United States of America 108 (4):1513-1518. doi:10.1073/pnas.1017351108
20. Maccallum I, Przybylski D, Gnerre S, Burton J, Shlyakhter I et al. (2009) ALLPATHS 2: small genomes assembled accurately and with high continuity from short paired reads. Genome Biol 10 (10):R103. doi:10.1186/gb-2009-10-10-r103
21. Zerbino DR, Birney E (2008) Velvet: Algorithms for de novo short read assembly using de Bruijn graphs. Genome research 18 (5):821-829. doi:10.1101/gr.074492.107
22. Zerbino DR, McEwen GK, Margulies EH, Birney E (2009) Pebble and rock band: heuristic resolution of repeats and scaffolding in the velvet short-read de novo assembler. PLoS One 4 (12):e8407. doi:10.1371/journal.pone.0008407
23. Simpson JT, Wong K, Jackman SD, Schein JE, Jones SJ et al. (2009) ABySS: a parallel assembler for short read sequence data. Genome research 19 (6):1117-1123. doi:10.1101/gr.089532.108
24. Li R, Zhu H, Ruan J, Qian W, Fang X et al. (2010) De novo assembly of human genomes with massively parallel short read sequencing. Genome research 20 (2):265-272. doi:10.1101/gr.097261.109
25. Li R, Li Y, Kristiansen K, Wang J (2008) SOAP: short oligonucleotide alignment program. Bioinformatics 24 (5):713-714. doi:10.1093/bioinformatics/btn025
26. Luo R, Liu B, Xie Y, Li Z, Huang W et al. (2012) SOAPdenovo2: an empirically improved memory-efficient short-read de novo assembler. GigaScience 1 (1):18. doi:10.1186/2047-217X-1-18
27. Ye C, Ma ZS, Cannon CH, Pop M, Yu DW (2012) Exploiting sparseness in de novo genome assembly. BMC bioinformatics 13 Suppl 6:S1. doi:10.1186/1471-2105-13-S6-S1
28. Warren RL, Sutton GG, Jones SJ, Holt RA (2007) Assembling millions of short DNA sequences using SSAKE. Bioinformatics 23 (4):500-501. doi: 10.1093/bioinformatics/btl629
29. Dohm JC, Lottaz C, Borodina T, Himmelbauer H (2007) SHARCGS, a fast and highly accurate short-read assembly algorithm for de novo genomic sequencing. Genome Res 17 (11):1697-1706. doi:gr.6435207
30. Jeck WR, Reinhardt JA, Baltrus DA, Hickenbotham MT, Magrini V et al. (2007) Extending assembly of short DNA sequences to handle error. Bioinformatics 23 (21):2942-2944. doi:10.1093/bioinformatics/btm451
31. Bryant DW, Jr., Wong WK, Mockler TC (2009) QSRA: a quality-value guided de novo short read assembler. BMC bioinformatics 10:69. doi:10.1186/1471-2105-10-69
32. Schmidt B, Sinha R, Beresford-Smith B, Puglisi SJ (2009) A fast hybrid short read fragment assembly algorithm. Bioinformatics 25 (17):2279-2280. doi:10.1093/bioinformatics/btp374
33. Wang Y, Yu Y, Pan B, Hao P, Li Y et al. (2012) Optimizing hybrid assembly of next-generation sequence data from Enterococcus faecium: a microbe with highly divergent genome. BMC Systems Biology 6 (3):1-13. doi:10.1186/1752-0509-6-s3-s21
34. Cerdeira LT, Carneiro AR, Ramos RTJ, de Almeida SS, D'Afonseca V et al. (2011) Rapid hybrid de novo assembly of a microbial genome using only short reads: Corynebacterium pseudotuberculosis I19 as a case study. Journal of Microbiological Methods 86 (2):218-223. doi:http://dx.doi.org/10.1016/j.mimet.2011.05.008
35. Martin J, Bruno VM, Fang Z, Meng X, Blow M et al. (2010) Rnnotator: an automated de novo transcriptome assembly pipeline from stranded RNA-Seq reads. BMC genomics 11:663. doi:10.1186/1471-2164-11-663
36. Sommer DD, Delcher AL, Salzberg SL, Pop M (2007) Minimus: a fast, lightweight genome assembler. BMC bioinformatics 8:64. doi:1471-2105-8-64

37. Schulz MH, Zerbino DR, Vingron M, Birney E (2012) Oases: robust de novo RNA-seq assembly across the dynamic range of expression levels. Bioinformatics 28 (8):1086-1092. doi:10.1093/bioinformatics/bts094
38. Grabherr MG, Haas BJ, Yassour M, Levin JZ, Thompson DA et al. (2011) Full-length transcriptome assembly from RNA-Seq data without a reference genome. Nat Biotech 29 (7):644-652.
39. Li H (2012) Exploring single-sample SNP and INDEL calling with whole-genome de novo assembly. Bioinformatics 28 (14):1838-1844. doi:10.1093/bioinformatics/bts280
40. Warren RL, Sutton GG, Jones SJM, Holt RA (2006) Assembling millions of short DNA sequences using SSAKE. Bioinformatics 23 (4):500-501. doi:10.1093/bioinformatics/btl629
41. Dohm JC, Lottaz C, Borodina T, Himmelbauer H (2007) SHARCGS, a fast and highly accurate short-read assembly algorithm for de novo genomic sequencing. Genome Res 17 (11):1697-1706. doi:10.1101/gr.6435207
42. Boisvert S, Laviolette F, Corbeil J (2010) Ray: simultaneous assembly of reads from a mix of high-throughput sequencing technologies. J Comput Biol 17 (11):1519-1533. doi:10.1089/cmb.2009.0238
43. Peng Y, Leung HM, Yiu SM, Chin FL (2010) IDBA—A Practical Iterative de Bruijn Graph De Novo Assembler. In: Berger B (ed) Research in Computational Molecular Biology, vol 6044. Lecture Notes in Computer Science. Springer Berlin Heidelberg, pp 426-440. doi:10.1007/978-3-642-12683-3_28
44. Peng Y, Leung HC, Yiu SM, Chin FY (2012) IDBA-UD: a de novo assembler for single-cell and metagenomic sequencing data with highly uneven depth. Bioinformatics 28 (11):1420-1428. doi:10.1093/bioinformatics/bts174
45. Peng Y, Leung HC, Yiu SM, Chin FY (2011) Meta-IDBA: a de Novo assembler for metagenomic data. Bioinformatics 27 (13):i94-101. doi:10.1093/bioinformatics/btr216
46. Chevreux B, Pfisterer T, Suhai S (2000) Automatic assembly and editing of genomic sequences. Paper presented at the Genomics and proteomics—functional and computational aspects, New York
47. Chevreux B, Wetter T, Suhai S (1999) Genome sequence assembly using trace signals and additional sequence information. Paper presented at the German Conference on Bioinformatics GCB'99, German
48. Zimin AV, Marcais G, Puiu D, Roberts M, Salzberg SL et al. (2013) The MaSuRCA genome assembler. Bioinformatics 29 (21):2669-2677. doi:10.1093/bioinformatics/btt476
49. Bankevich A, Nurk S, Antipov D, Gurevich AA, Dvorkin M et al. (2012) SPAdes: a new genome assembly algorithm and its applications to single-cell sequencing. J Comput Biol 19 (5):455-477. doi:10.1089/cmb.2012.0021
50. Chikhi R, Rizk G (2012) Space-Efficient and Exact de Bruijn Graph Representation Based on a Bloom Filter. In: Raphael B, Tang J (eds) Algorithms in Bioinformatics, vol 7534. Lecture Notes in Computer Science. Springer Berlin Heidelberg, pp 236-248. doi:10.1007/978-3-642-33122-0_19

Concluding Remarks

The field of next-generation sequencing is a multidisciplinary field that requires a firm background in molecular biology and computer science. Most individuals in the field tend to have specialized in one of these areas before obtaining sufficient knowledge in the other. With this in mind, we provided an introduction to both these fields (Chaps. 1 and 2) at the beginning of this book. This overview would greatly assist readers with limited background in either of these areas to understand the concepts discussed in this book and other next-generation sequencing literature.

With the rapid development of the next-generation sequencing field in the last few years, the platform has experienced significant advancements in all areas. Sequencing costs, time, and effort have dropped remarkably to allow more and more laboratories to adopt this indispensable technology. Furthermore, efforts dedicated to the automation of library preparation have provided a drastic reduction in error sources and allowed the development of compact sequencing instruments that combine the library preparation, sequencing, and data analysis functions into a single bench-top machine.

Parallel to the developments observed in sequencing technologies, the features of the resulting sequences themselves have experienced outstanding improvement. While initial reads comprised of lengths up to 35 bp, next-generation sequencing has allowed the development of instruments that output reads with lengths of 700 bp or even longer. The most recent sequencing instruments, considered to be third-generation sequencing technology, have expanded upon this number to produce reads exceeding 3,000 bp. The availability of increased read lengths has had the positive effect of reducing the amount of data analysis and computational resources required for assembling the sequence.

Despite the notable developments and advancements in next-generation sequencing technologies, the methodology continues to face certain challenges. While expenses related to sequencing have dropped dramatically to allow affordability for the average laboratory, the costs involved in establishing a new sequencing facility remain beyond the reach of most institutions. Furthermore, the establishment and maintenance of sequencing facilities in developing countries remain a major challenge.

S. El-Metwally et al., *Next Generation Sequencing Technologies and Challenges in Sequence Assembly*, SpringerBriefs in Systems Biology 7, DOI 10.1007/978-1-4939-0715-1, © The Authors 2014

The size of the read length and significant error rates are two additional challenges that need further development and improvement in relation to next-generation sequencing. However, read lengths have already experienced considerable increases with the utilization of recent sequencing instruments, and ongoing developments promise to augment these values even further. In contrast, the improvement in error rates has not been quite as substantial in these recent instruments. The second part of this book (Chaps. 3–7) has provided an overview of next-generation sequencing methods, platforms, applications, developments, and challenges. Here, we tried not to emphasize on current technical and procedural challenges as it is our contention that most of these obstacles will gradually be alleviated given the history and rapid pace of development in the field. Therefore, other publications such as review articles may be more suitable for these purposes.

In addition to next-generation processes linked with the attainment of sequence reads, a major challenge in the form of attempting to assemble these structures into full genomes or chromosomes also awaits. Steps and associated challenges related to solving this sequence assembly puzzle in the next-generation environment involve a series of interleaved stages that are discussed in detail in the third part of this book (Chaps. 8 and 9). The assembly process encompasses a complex five-phase process that includes the error correction phase, the graph construction phase, the graph simplification phase, the scaffolding phase, and the assembly assessment phase. We have devoted an independent chapter (Chap. 10) to the assembly assessment phase where we reviewed currently available methods and tools for the evaluation of the quality of the assembled sequence. To the best of our knowledge, this is the first attempt to review this topical subject.

The construction of an assembler in the next-generation environment is not a trivial task for developers in accordance with different stakeholders' requirements. Non-expert users require more responsive assemblers with user-friendly interfaces and easy installation packages. On the other hand, biologists prefer assemblers with different visualization, statistical and assessment tools to perform their analysis efficiently. Hence, developers have attempted to fulfill these requirements with available computational resources to manage the ever-increasing data-generation rates from modern sequencing technologies. In Chap. 11, we have reviewed over 30 tools for next-generation genome, transcriptome, and metagenome sequence assembly.

We believe that this book provides the reader with useful insight into the history, basic principles, methods, applications, and challenges related to the next-generation sequencing field. In reviewing the above topics, the authors of this book eagerly anticipate the future scientific and technological outcomes of this exciting field.